T0073124

Research Methodologies for Beginners

Kitsakorn Locharoenrat

PAN STANFORD PUBLISHING

Published by

Pan Stanford Publishing Pte. Ltd.
Penthouse Level, Suntec Tower 3
8 Temasek Boulevard
Singapore 038988

Email: editorial@panstanford.com
Web: www.panstanford.com

British Library Cataloguing-in-Publication Data
A catalogue record for this book is available from the British Library.

ISBN 978-981-4745-39-0 (Hardcover)
ISBN 978-1-315-36456-8 (eBook)

Printed in Canada

Contents

Research
Methodologies
for Beginners

Preface

In this textbook, I introduce the general viewpoints on the research methodology adopted in the science and engineering fields of study. I present an overview of the technical and professional communication required for journal publication, which is necessary for the survival of independent beginners. I refer to my own academic papers to explain the related theoretical and practical concepts that I have developed in the past 10 years.

There are many practice activities that help beginners gain confidence in communication, but they require some simple skills. I have introduced these skills gradually and methodically, in measured amounts and in a logical order. Chapter 1 introduces a general viewpoint on scientific research methodology to explore new information and analyze the cause-and-effect relation for specific problems. This is a systematic way of finding out useful data on science and engineering issues. Chapter 2 explains how to select a topic for study from the defined problem. The sources for the selection of the topic for a paper come from basic research, applied research, or both. Chapter 3 suggests how to write a good abstract and conclusion and Chapter 4 the ways to search all relevant literature to get readers' attention and interest. Chapter 5 shows how to present the detailed methodology so that readers can replicate the work to check the study. Chapter 6 demonstrates how to accurately collect data and obtain results for readers to follow and how to interpret the obtained results. Chapter 7 discusses research ethics based on the practice standards recommended by the Committee on Publication Ethics. The papers not conforming to these standards are removed if plagiarism or duplicate content is discovered at any time, even after the publication. Chapter 8 focuses on the presentation format, which will help readers share the idea of their work and/or get financial support from funding agencies.

The problem sets provided in each chapter examine readers' understanding of each concept. After reading this book, readers will be able to formulate a specific research topic, research questions, and hypotheses. They will be well equipped to conduct a literature review relevant to the research topic and develop an applicable research methodology. Finally, they should be able to write and present their research, outlining the key elements of the project.

I would like to thank King Mongkut's Institute of Technology Ladkrabang (KMITL), Thailand, especially the Faculty of Science, Department of Physics, for the support and cooperation provided for writing this book. I would appreciate the readers' comments and suggestions for improving the book further.

Kitsakorn Locharoenrat

Fall 2016

Chapter 1

Introduction

In this chapter, I introduce a general viewpoint of scientific research methodology to help you explore new information and analyze the relation between causes and effects of specific problems. This is a systematic way to find out useful data on scientific and engineering issues.

1.1 Overview

For writing scientific and engineering papers, first you have to determine which field you are interested in. You should find the journals that publish papers in your field of study. Then, you must read about 5–10 papers published by the journals in the past 5–10 years to be familiar with the type of publication, writing style, and format. You can access the website of publishers (Table 1.1) and submit your manuscript according to author guidelines. Papers can be submitted to an independent journal with an ISSN or to a special issue of conference proceedings having an ISBN. After the manuscript is submitted, you have to address the comments made by several reviewers. The paper is accepted and published after it is finally revised.

Research Methodologies for Beginners
Kitsakorn Locharoenrat
Copyright © 2017 Pan Stanford Publishing Pte. Ltd.
ISBN 978-981-4745-39-0 (Hardcover), 978-1-315-36456-8 (eBook)
www.panstanford.com

Table 1.1   Global ranks of publishers of some scholarly journals

No.	Name of publisher	Rank
1.	Elsevier	1
2.	Springer	2
3.	Wiley	3
4.	Taylor & Francis	4

1.2   Impact Factor

You should not send your paper to predatory journals and publishers, as per Jeffrey Beall's list. You should send it to authentic journals having an *impact factor*. The impact factor is considered the number "one" ranking value for authentic journals and has become a substantial part of any journal development discussion.

Eugene Garfield introduced it in the late 1950s. The yearly developments in Impact Factor are reported in the Thomson Reuters. Impact factor is a benchmark for a journal's reputation and reflects how frequently peer-reviewed journals are cited by other researchers in a particular year. It will help you to evaluate the journal's relative importance, especially when you would like to compare your work with others in the same field by checking a link at the ISI Web of Knowledge and Journal Citation Reports (JCR) (see Table 1.2).

Table 1.2   A JCR summary

Abbreviated journal title	ISSN	Total citation	Impact factor
ACTA PHYS POL A	0587–4246	2080	0.604
ACTA PHYS SLOVACA	0323–0465	273	2.000
ADV COND MATTER PHYS	1687–8108	200	0.862
ADV MATER SCI ENG	1687–8434	428	0.744
CHEMIJA	0235–7216	124	0.357
CHINESE J CHEM ENG	1004–9541	1572	0.872
CHINESE J PHYS	0577–9073	433	0.413
CROATICA CHEMICA ACTA	0011–1643	1026	0.556

Abbreviated journal title	ISSN	Total citation	Impact factor
ENVIRON ENG MANAG J	1582–9596	1137	1.258
J CHEM	0973–4945	998	0.696
J NANO RES-SW	1662–5250	248	0.564
J SPECTROSC	2314–4920	80	0.538
MACED J CHEM ENG	1857–5552	86	0.533
MATER TEHNOL	1580–2949	324	0.548
MATER TRANS	1345–9678	6792	0.679
MATERIA-BRAZIL	1517–7076	23	0.074
OPT REV	1340–6000	658	0.546
POLYM-KOREA	0379–153X	325	0.433
REV METAL MADRID	0034–8570	152	0.355
REV MEX FIS	0035–001X	526	0.328
ROM J PHYS	1221–146X	450	0.745
SCIENCEASIA	1513–1874	278	0.347
TURK J BIOCHEM	0250–4685	128	0.173
TURK J CHEM	1300–0527	1230	1.176
UKR J PHYS OPT	1609–1833	90	0.558

1.3 Research Criteria

An authentic paper is generally composed of the following criteria:

1. *Necessity of research.* If your answer to one of the following questions is YES, your published paper is required:
   - Should the area be addressed because it has been neglected?
   - Would the area fill a gap in current scientific knowledge?
2. *Significance of research.* If your answer to one of the following questions is YES, your published paper is important:
   - Does your paper advance understanding?
   - Does your paper make a useful contribution?
   - Can you show a specific instance of the possible useful contribution?
   - Are your findings original?

3. *Relevance of research.* If your answer to one of the following questions is YES, this is an appropriate place for your possible publication:
   - Is your research of interest to the journal audience?
   - Is this topic relevant to the publication?

4. *Clarity of research.* If your answer to one of the following questions is YES, your paper is easily acceptable:
   - Is your topic clearly mentioned?
   - Does your paper follow through by addressing this topic consistently and cogently?

5. *Relation to literature.* If your answer to one of the following questions is YES, your paper is reliable:
   - Does your paper show an adequate understanding of the current literature in the field?
   - Does your paper connect with the literature in a way which might be useful to the development of our understanding in the area it addresses?

6. *Reference of research.* If your answer to one of the following questions is YES, your findings from other research are very useful:
   - Does your paper use theory in a meaningful way?
   - Does your paper develop or employ theoretical concepts in such a way as to make plausible generalizations?
   - Does your paper develop new theory?
   - Does your paper apply theory to practice?

7. *Methodology of research.* If your answer to one of the following questions is YES, your research design and data collection methods are suitable:
   - Has the research, or equivalent intellectual work upon which the paper is based, been well designed?
   - Does the paper show adequate use of evidence, informational input, or other intellectual raw materials in support of its case?

8. *Data analysis.* If your answer to one of the following questions is YES, you can use data correctly:
   - Can you correctly use data analysis techniques?

- Can you make right interpretations from the data collected?
- Have the data been used effectively to advance the themes that the paper sets out to address?

9. *Critical qualities.* If your answer to one of the following questions is YES, your paper has good quality:
   - Does your paper demonstrate a critical self-awareness of the author's own perspectives and interests?
   - Does your paper show awareness of the possibility of alternative or competing perspectives, such as other theoretical or intellectual perspectives?
   - Does your paper show an awareness of the practical implications of the ideas it is advancing?

10. *Clarity of conclusions.* If your answer to one of the following questions is YES, your paper is cohesive:
    - Are the conclusions of the paper clearly stated?
    - Do the conclusions adequately tie together the other elements of the paper (such as theory, data, and critical perspectives)?

11. *Quality of communication.* If your answer to one of the following questions is YES, your paper is well written:
    - Does your paper clearly express its case, measured against the technical language of the field and the reading capacities of an academic, tertiary student, and professional readership?
    - What is the standard of writing, including spelling and grammar?

If you say YES for almost every criteria, this means that your authentic paper possibly makes it clear how the findings advance understanding of the issue under study. In addition, your paper provides an important critical and/or analytical insight that contributes something new to the field of higher education. Moreover, your issue/problem is well situated in appropriate literature. Furthermore, your paper demonstrates methodological soundness. Last, your conclusion is well supported and persuasively argued. Your paper is also succinct and coherent.

1.4 Format of Paper

Generally, there are many types of the authentic papers. Herein we focus on research paper and review paper, which many researchers usually publish for their experimental and/or theoretical works.

1. *Research paper.* A research paper can be quantitative, qualitative, or mixed methods. This paper shows important and original findings.

2. *Review paper.* A review paper is usually commissioned by editors. This paper raises technical or scientific questions about the published work.

   A general set of sequential components of the research and review papers should have the following details (see also Table 1.3):

   - *Title.* The title length is usually around 5–20 words. Context must be short and specific.
   - *Abstract.* Abstract is a brief summary of about 75–300 words. You can use the suitable tense in the sentence. You should avoid citations, tables, equations, figures, and references.
   - *Keywords or PACS numbers.* You must provide at least five keywords or the Physics and Astronomy Classification Scheme (PACS) numbers so that these are useful to select reviewers who are familiar with your papers. PACS was developed by the American Institute of Physics. It is used to identify the fields in physics, astronomy, and related sciences.
   - *Introduction.* The introduction must show a statement of the problem, significance of the study, and objective.
   - *Methodology.* The methodology identifies the type of experimental study and general design and construction. We normally use the past tense but the active voice in this section.
   - *Results and discussion.* Data analysis must be accurate. This must be based on the research questions you have formulated. Results and discussions can be either in the same section or in separate sections. You should show and

compare your findings with other studies. In addition, you should purpose your own suggestions.

- *Conclusion.* The content can be similar to the introduction plus abstract sections, but you can just restate it.
- *Acknowledgment* (if any). You mention the financial support and thank your college at this section.
- *References.* You must show the references cited in the paper using the reference format according to the publisher's format.

Table 1.3   Sequential components of the paper

1.	Title *You write a few tentative titles and select the best one later.*
2.	Abstract *You write an abstract when the whole content of the document is completed.*
3.	Keywords and/or PACS numbers
4.1	State a conceptual framework *You show a block diagram and connect your variables to the theories.*
4.2	Literature review based on the research problem to solve/explain/ test and develop. *You provide the evidence of the existence of the problem.* *You cite the previous studies that are inconclusive or contradictory.* *You prove that there is lack of studies in this area and it is important to fill this gap.*
4.3	Objective *Based on the research problem, state what your study intends to do and what the significance of this study is.*
5.	Methodology *What the sample would comprise, how sampling would be done, and what instruments would be used for sampling?*
6.1	Results *You state the results based on each research problem.*
6.2	Discussion *You compare findings with the findings from previous studies reviewed in the literature review. You give reasons why your findings support/do not support the previous findings.*
7.	Conclusion
8.	References

1.4.1 Example of Paper Template

Recent Trend of Solar Tracking System for Electric Power Conversion

Sarai Lekchaum,[1] Kitsakorn Locharoenrat[1, a*]

[1]Department of Physics, Faculty of Science, King Mongkut's Institute of Technology Ladkrabang, Bangkok 10520, Thailand

[a]kitsakorn.lo@kmitl.ac.th

Abstract. In this article,

....
....
....
....
....
....
....
....

Keywords: altitude, analemma, azimuth, thermal energy, solar tracker.

Introduction

The solar trackers mostly use a sensor module [1–5] or the sun position control systems [6, 7], however, there are some benefits and drawbacks for these systems.

....
....
....
....
....
....
....
....
....
....

Design and Fabrication

In principle, we calculated the position of the sun in the horizon system in terms of altitude and azimuth angles as shown in Fig. 1 [12].

....
....
....

....
....

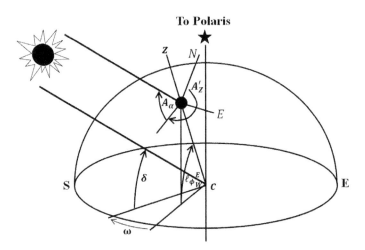

Figure 1 Position of the sun in the horizon coordination system. C is earth's center.

....
....
....
....
....

Table 1 Coefficient A_k and B_k

k	A_k [hr]	B_k [hr]
0	2.0870×10^{-4}	0
1	9.2869×10^3	-1.224×10^{-1}
2	-5.2258×10^{-2}	-1.5698×10^{-1}
3	-1.3077×10^{-3}	-5.1602×10^{-3}
4	-2.1867×10^{-3}	-2.9823×10^{-3}
5	-1.5100×10^{-4}	-2.3463×10^{-4}

Results and Discussion

The experiment is carried out at King Mongkut's Institute of Technology Ladkrabang, Bangkok, Thailand located at latitude ($\ell_{\phi W}^E$ = 13°43′ N) and longitude ($\ell_{\lambda w}^E$ = 100°47′ E).

....
....
....
....
....

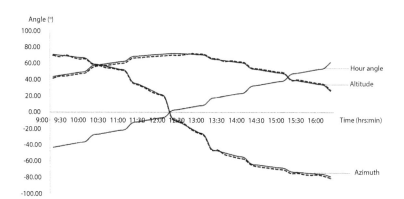

Figure 2 Altitude and azimuth angles from calculation and experiment (dashed and solid lines stand for experimental and calculation data, respectively), as well as hour angle on 13-Mar-2015 between 09:00 and 16:00.

....
....
....
....
....

Conclusion

We have designed and constructed the solar tracking system to determine the position of the solar motion at King Mongkut's Institute of Technology Ladkrabang, Bangkok, Thailand with latitude of 13°43'N and longitude of 100°47'E for the year 2015.

....
....
....
....
....

Acknowledgments

The authors would like to thank

....

References

[1] M. T. A. Khan, S. M. S. Tanzil, R. Rahman, and S. M. S. Alam, Design and construction of an automatic solar tracking system, *6th Int. Conf. Elec. Compt. Eng. (ICECE)* (2010), pp. 326–329.

[2] U. K. Okpeki and S. O. Otuagoma, Design and construction of a bi–directional solar tracking, *Int. J. Eng. Sci.*, 2(5) (2013), pp. 32–38.

[3] Y. C. Park and Y. H. Kang, Design and implementation of two axes sun tracking system for the parabolic dish concentrator, *ISES2001 Solar World Congress* (2001), pp. 749–760.

[4] X. Jin, G. Xu, R. Zhou, X. Luo, and Y. Quan, A sun tracking system design for a large dish solar concentrator, *Clean Coal Energy*, 2(2B) (2013), pp. 16–30.

[5] P. Roth, A. Georgiev, and H. Boudinov, Design and construction of a system for sun-tracking, *Renew. Energy*, 29(3) (2004), pp. 393–402.

[6] Y. Rizal, S. H. Wibowo, and Feriyadi, Application of solar position algorithm for sun-tracking system, *Energy. Proc.*, 32 (2013), pp. 160–165.

[7] M. Mirdanies, Astronomy algorithm simulation for two degrees of freedom of solar tracking mechanism using C language, *Energy Proc.*, 68 (2015), pp. 60–67.

[8] L. Morison, *Introduction to Astronomy and Cosmology*, John Wiley & Sons, London, 2008.

[9] S. Ray, Calculation of sun position and tracking the path of sun for a particular geographical location, *Int. J. Emer. Technol. Adv. Eng.*, 2(9) (2012), pp. 81–84.

[10] W. B. Stine and R. W. Harrigan, *Solar Energy Fundamentals and Design: With Computer Applications*, John Wiley & Sons, New York, 1985.

[11] G. Prinsloo and R. Dobson, *Solar Tracking*, Prinsloo, South Africa, 2014.

[12] J. A. Duffie and W. A. Beckman, *Solar Engineering of Thermal Processes*, John Wiley, New York, 2013.

[13] N. D. Kaushika and K. S. Reddy, Performance of a low cost solar paraboloidal dish steam generating system, *Energy Conversion Mgt.*, 41 (2000), pp. 713–726.

[14] R. Y. Nuwayhid, F. Mrad, and R. Abu-Said, The realization of a simple solar tracking concentrator for university research applications, *Renew. Energy*, 24(2) (2001), pp. 207–222.

[15] S. Pairoj, A parabolic solar concentrator with different receiver materials, Master Thesis, King Mongkut's Institute of Technology Ladkrabang, Thailand.

1.5   Status of Report

After your paper submission, you would have to deal with comments from reviewers. You can face any one of the three scenarios: outright rejection, major corrections, and minor corrections. Very rarely, there is outright acceptance without corrections.

1. *Outright rejection.* You must calm down so that you can think better. First you must go through the comments and reasons provided. If valid, you consider major revisions. If your major revisions involve collecting data, you must consider the time and money needed. If you just require to reanalyze the data, rewriting the paper based on the comments is easier.

2. *Major corrections.* There is still some hope. If it is going back to collecting data, you must consider time and money involved, subject availability. If it is just reanalyzing the data, you seek the statistician's help and try to incorporate the reviewers' comments. If it is just a conceptualization issue, it is easier to solve. You just look at some examples of similar papers and how the problems are conceptualized.

3. *Minor corrections.* The editor considers your paper to be almost accepted. You follow the comments closely and do not unnecessarily argue with the editor. The editor makes a final decision based on the comments of the reviewers. You revise the paper within the time allocated and send it to a language expert to check and then send the paper a few days before the deadline.

1.6   Problems

Pb.1. What is the importance of research and why people conduct research?

Pb.2. Discuss the differences between research paper and review paper.

Pb.3. Explain the structure of a report.

Pb.4. Give a few examples of predatory journals and publishers.

Pb.5. Give a few examples of authentic journals, including the ISI Impact Factor, ISSN, and total cites in your field of study.

Pb.6. Write the steps involved in writing a paper from the beginning until the publication.

Chapter 2

Selection of Topic

In this chapter, I will explain how to select a topic for your study. The context of what your study plans to do must be short and clear. You should mention all the key variables that will be investigated. Sources for the paper topic should be selected based on basic research (i.e., theory of your field of interest) or applied research (i.e., daily problems, recent trends of technologies) or both.

2.1   Topic of the Study

The topic must be specific and focused. The title length of the topic is usually around 5–20 words (see Table 2.1). The dependent or main variables may be mentioned here.

2.2   Basic Research

In basic research, or theoretical research, you will study the basic concepts and reasons for the phenomena relating to pure science or engineering. Basic research does not lead to any applications; it is original or basic in character. It offers a logical, systematic, and deep understanding of the scientific problem. It may help you in finding new developments related to a specific topic. Research on

Research Methodologies for Beginners
Kitsakorn Locharoenrat
Copyright © 2017 Pan Stanford Publishing Pte. Ltd.
ISBN 978-981-4745-39-0 (Hardcover), 978-1-315-36456-8 (eBook)
www.panstanford.com

improving a theory is also referred to as basic research. For example, the quantum theory is applicable to a system provided the system satisfies certain specific conditions. A modification of the theory is applicable to a general situation. The output of basic research will help you to venture into applied research.

Table 2.1 Examples of paper titles and word count

No.	Paper title	Word count
1.	Shadow Deposition of Copper Nanowires on the Facted NaCl(110) Template	10
2.	Phenomenological Studies of Optical Properties of Cu Nanowires	8
3.	Rotational Anisotropy* in Second Harmonic Intensity from Copper Nanowire Arrays on the NaCl(110) Substrates	14
4.	Second Harmonic Spectroscopy of Copper Nanowire Arrays on the (110) Faceted Faces of NaCl Crystals	16
5.	Quantification of Fluorescence Target in Tissue Phantoms by Time-Domain Diffuse Optical Tomography with Phantoms—Total-Light Approach	16
6.	Time-Domain Fluorescence Diffuse Optical Tomography for Live Animals by Total-Light Algorithm	11
7.	Compact Optical Delay Line for Long-Range Scanning	7
8.	Optical Delay Line for Rapid Scanning Low-Coherence Reflectometer	8
9.	Nitrogen Concentrations* on Structural and Optical Properties of Aluminum Nitride Films Deposited by Reactive RF-Magnetron Sputtering	16
10.	Structure* and Piezoelectric Properties of Aluminum Nitride Thin Films on the Quartz Substrates Deposited by Reactive RF-Magnetron Sputtering	18
11.	Construction of the Optical Delay Line for the Optical Coherence Tomography	11
12.	Investigation of Temporal Profiles at the Symmetrical Points of the Target in Tissue Phantoms by Time-Resolved Fluorescence Diffuse Optical Tomography	20
13.	Preparations and Field Emission from Tungsten Nanotips	7

No.	Paper title	Word count
14.	Optical Property of Indocyanine Green in a Tissue Model	9
15.	Plasmonic Properties of Gold-Palladium Core-Shell Nanorods	6
16.	Optical Studies of Zinc Oxide Nanoparticles and Their Biomedical Application	10
17.	Enhancement of Fluorescence in Inorganic Dyes by Metallic Nanostructured Surfaces	10
18.	Preparation and Metal Removal from Chitosan/PEG Blend	7
19.	Recent Trend of Solar Tracking System for Electric Power Conversion	10
20.	Fluorescent Dye Doped ZnO Nanoparticles for Photovoltaic Application	8
21.	Demonstration of Gamma-Type Stirling Engine's Performance onto Parabolic Dish of Solar Concentrator	12
22.	Design and Construction of New Hybrid Solar Tracking System	9
23.	Generation of 2nd-Harmonic Light from Noncentrosymmetric Material	7

*key variables

2.2.1 Some Examples of Basic Research

1. General fields of physics
- Communication, education, history, and philosophy
- Mathematical methods in physics
- Quantum mechanics, field theories, and special relativity
- General relativity and gravitation
- Statistical physics, thermodynamics, and nonlinear dynamical systems
- Metrology, measurements, and laboratory procedures
- Instruments, apparatus, and components common to several branches of physics and astronomy

2. Elementary particles
- General theory of fields and particles

- Specific theories and interaction models
- Specific reactions and phenomenology
- Properties of specific particles

3. **Nuclear physics**
   - Nuclear structure radioactive decay and in-beam spectroscopy
   - Nuclear reactions
   - Nuclear astrophysics
   - Properties of specific nuclei listed by mass ranges
   - Nuclear engineering and nuclear power studies
   - Experimental methods and instrumentation for elementary-particle and nuclear physics

4. **Atomic and molecular physics**
   - Electronic structure of atoms and molecules
   - Atomic properties and interactions with photons
   - Molecular properties and interactions with photons
   - Atomic and molecular collision processes and interactions
   - Exotic atoms and molecules
   - Mechanical control of atoms, molecules, and ions

5. **Condensed matter**
   - Structure of solids and liquids
     Mechanical and acoustical properties of condensed matter
   - Lattice dynamics
   - Equations of state, phase equilibrium, and phase transitions
   - Thermal properties of condensed matter
   - Non-electronic transport properties of condensed matter
   - Quantum fluids and solids
   - Surfaces and interfaces; thin films and nanosystems
   - Electronic structure of bulk materials
   - Electronic transport in condensed matter
   - Electronic structure and electrical properties of surfaces, interfaces, thin films, and low-dimensional structures
   - Superconductivity

- Magnetic properties and materials
- Magnetic resonances and relaxations in condensed matter
- Dielectrics, piezoelectrics, and ferroelectrics and their properties
- Optical properties, condensed matter spectroscopy, and other interactions of radiation and particles with condensed matter
- Electron and ion emission by liquids and solids

6. **Geophysics and astronomy**
   - Solid earth physics
   - Hydrospheric and atmospheric geophysics
   - Geophysical observations, instrumentation, and techniques
   - Physics of the ionosphere and magnetosphere
   - Fundamental astronomy and astrophysics; instrumentation, techniques, and astronomical observations
   - Solar system
   - Stars and the universe

7. **Related areas of science and engineering**
   - Strength of materials
     Rheology
   - Momentum, heat, and mass transfer
   - Renewable energy resources
   - Fluid dynamics

2.3   Applied Research

Applied research is the outcome of direct or indirect applications. Such research is of practical use to the current activity. For example, researches on environmental problems have immediate use. Applied research is concerned with real-life issues such as research on increasing the efficiency of solar cells, increasing the quantum yield of materials, preparing drug delivery to cancer cells, etc. Obviously, these have immediate potential applications.

2.3.1   Some Examples of Applied Research

- Physical chemistry
- Chemical physics
- Biological physics
- Medical physics
- Biomedical physics

2.4   Problems

Pb.1. Identify some sources of topic selection.

Pb.2. Select some topic of current interest or recent trends.

Pb.3. Distinguish between theory and experiment.

Pb.4. Write a note on the importance of theory in basic and applied researches.

Pb.5. Discuss the relation between qualitative and quantitative methods.

Pb.6. Discuss the following paper titles for suitability.

  (i)   Self-Organized Copper Nanowires Studied by Second Harmonic Spectroscopy

  (ii) Field Enhancement in Arrays of Copper Nanowires Investigated by the Finite-Difference Time-Domain Method

 (iii) Demonstration of Confocal Sum Frequency Microscopy

 (iv)   Construction of an Optical Sum Frequency Microscope with Confocal Optics

  (v)   Characterization, Optical and Theoretical Investigation of Arrays of the Metallic Nanowires Fabricated by a Shadow Deposition Method

 (vi)   Nonlinear Optical Properties of Controlled Fabrication of Copper Nanowires by a Shadow Deposition

(vii)   Copper Nanowires on NaCl(110) Template

(viii) Investigation of Temporal Profiles at the Symmetrical Points of the Target in Tissue Phantoms by Time-Resolved Fluorescence Diffuse Optical Tomography

 (ix)   Second-Order Nonlinear Optical Response of Metal Nanostructures

(x) Mechanism of Resonant Enhancement of Gold- and Copper-Nanowires Arrays, Advanced Materials Research
(xi) Recent Advances in Nanomaterial Fabrication
(xii) Second Harmonic Generation Based on Strong Field Enhancement in Metallic Nanostructured Surface

Chapter 3

Clarity of Abstract and Conclusion

In this chapter, I will explain how to write abstracts and conclusions. An abstract is a description of the entire document and is the first paragraph of a paper. Conclusion is the last paragraph of the paper. It is like the final chord in the song. It makes the listeners/readers understand that the piece is completed.

3.1 Abstract

You can write an abstract only when the whole document is complete. When you write an abstract, you should keep in mind two purposes: giving readers useful information included in the document and helping them to get basic information about the document.

Some general guidelines for writing an abstract are as follows:

1. An abstract is a brief summary of about 75–300 words. It should be short and should contain concise sentences in one paragraph.

2. An informative abstract should have the following order:

    - You make a statement of purpose indicating the situation, problem, or issue that will be studied and why the research should be undertaken (i.e., objective or hypothesis).

Research Methodologies for Beginners
Kitsakorn Locharoenrat
Copyright © 2017 Pan Stanford Publishing Pte. Ltd.
ISBN 978-981-4745-39-0 (Hardcover), 978-1-315-36456-8 (eBook)
www.panstanford.com

- You indicate the variables involved and how these will be measured (i.e., technique or method).
- You indicate what data analyses will be undertaken and show the results that will explain or solve the issue. (i.e., observation or evaluation).
- You make a conclusion and provide recommendation for further research.

3. While writing an abstract, you should avoid some tactics. You should not mention the names of authors. You should avoid tables, equations, graphs, references, or other citations.

3.1.1 Some Examples of Informative Abstract

Abstract: Example 1 (115 words)

We have studied microstructures and optical properties of self-organized arrays of copper nanowires. Shadow deposition onto (110) faceted faces of NaCl crystal has been demonstrated as a simple technique for fabricating wide-area arrays suitable for both transmission electron microscopy and optical measurements. Optical absorption spectra of the thinner nanowires exhibited stronger anisotropic absorption than the hicker nanowires. Absorption maxima are located at lower photon energy when the incident field is parallel to the wire axes than they are perpendicular to each other. They also shift to lower energy when the widths of the nanowires are increased. The Maxwell–Garnett model was found to be inappropriate to explain the present results except the case of non-retardation limit.

Abstract: Example 2 (102 words)

We have reported on the experimental observation of surface plasmon resonance in Cu nanowires fabricated by the shadow deposition method. When the incident light is polarized perpendicular to the wire axes, plasmon maxima appeared at about 2.3 eV in the absorption spectra. Plasmon resonance appeared at lower photon energy when the incident light is polarized parallel to the wire axes. Resonance peaks move to lower energy when the nanowire widths are increased. We have found that finite-difference time-domain (FDTD) simulation gives better results than the Maxwell–Garnett model in explaining the relation between the light polarization and the energies of the observed absorption maxima.

Abstract: Example 3 (78 words)

We have measured the azimuthal angle dependence of the second harmonic (SH) intensity from Cu nanowires on the faceted NaCl(110) substrates in air at the fundamental photon energy of 1.17 eV. The SH intensity patterns showed two main lobes for p-in/p-out, s-in/p-out, and s-in/s-out polarization configurations. From the results of the experiment and the pattern analysis, we have found that the observed SH light is enhanced by the electric field components along the substrate normal.

Abstract: Example 4 (90 words)

We have investigated the spectral dependence of the optical second harmonic signal from Cu nanowires deposited on the NaCl(110) faceted templates as a function of the second harmonic photon energies from 2.4 to 4.6 eV. The second harmonic response exhibited a peaked resonance near the second harmonic two-photon energy of 4.4 eV for the p-in/p-out and s-in/p-out polarization combinations. At this photon energy, the second harmonic response due to the resonant coupling between the fundamental field and the local plasmons in the wires is suggested to be dominant.

Abstract: Example 5 (75 words)

The applicability of total-light algorithm has been tested with tissue phantoms. It was experimentally confirmed that this algorithm can construct the background image in the absence of the fluorescence target from temporal profiles of the excitation and the fluorescence in the presence of the target. Then, the absorption image of the fluorescent target could be successfully obtained. This method can be useful in the quantification of the fluorophores in inhomogeneous material, such as whole tissues.

Abstract: Example 6 (98 words)

We have reported the first trial image reconstruction of an implanted fluorescent target into a live rat abdomen. We use a simplified algorithm for fluorescence diffuse optical tomography, so-called the total-light algorithm to obtain the absorption image of the target from the measured mean-transit time. We reconstructed two absorption images with and without a fluorescence target. It is difficult to identify something in the absorption images. However, the difference image between the two images highlights the target. This suggests that our algorithm is robust to the artifacts in the images in the real situation of in vivo measurements.

Abstract: Example 7 (161 words)

We have proposed and demonstrated an optical delay line composed of all reflective components for long-range scanning without walk-off problem. The optical delay line consists of a retro-reflector, an inclined reflection mirror, and a scanning mirror. The size of the optical delay line is within 2 cm × 2 cm, and the scanning range can reach 2.9 mm, when the beam is incident at the pivot of the scanning mirror and the vibration angle of the scanning mirror is 9.6°. The scanning range can be further increased when the pivot of the scanning mirror is laterally deviated from the incident beam. The optical delay line possesses the advantages that it is compact, easy to fabricate, and can perform rapid scanning in large scanning range without walk-off problem. The optical delay line was demonstrated with a low-coherence reflectometer where the scanning rate was 400 Hz. A higher scanning rate can be achieved when a scanning component with higher scanning rate is applied.

Abstract: Example 8 (118 words)

We have fabricated aluminum nitride films on quartz substrates using radio frequency reactive magnetron sputtering method. The conditions of the films have been performed under different concentration ratios between nitrogen and argon. We have found that all obtained films were transparent in visible wavelength. By using X-ray diffraction technique, it was found that the (002), (102), and (103) orientations were shown in X-ray diffraction patterns. The (002) orientation was dominant when nitrogen concentration was at 40%. On the other hand, the refractive index and optical band gap energy of the films were determined as a function of nitrogen concentration. We have found that the refractive index weakly depended on CN, while optical band gap energy did not.

Abstract: Example 9 (153 words)

We have presented the effect of nitrogen concentration on aluminum nitride bonding formation, structure and morphology of the aluminum nitride films. The films on the unheated substrates were deposited by radio frequency reactive magnetron sputtering technology using an aluminum target under argon/nitrogen mixture atmosphere. The FTIR and Raman spectra of the films confirmed their absorption bands corresponding to E1 (TO), A1 (TO), and E2 (high) vibration modes of the infrared active aluminum nitride bonding. The crystallographic orientation of the films was optimized under nitrogen concentration of 40%. The cross-sectional FE-SEM image of the film under this condition showed the columnar structure. The dense columnar grains were uniformly observed on the films surface under all nitrogen concentration, except for nitrogen concentration of 20%. The bulk resistivity and piezoelectric property were investigated via the metal–insulator–metal structures. The results showed that the resistivity was in a range of 10^{14}–10^{15} Ωcm while the effective piezoelectric coefficient was 11.03 pm/V.

Abstract: Example 10 (81 words)

We have constructed the compact optical delay line for the axial scanning of time-domain optical coherence tomography. The delay line contains the retro-reflector, and the inclined reflection mirror, as well as the scanning mirror. This delay line is performed by the low-coherence reflectometer with a scanning speed of 400 Hz. The dimension of the delay line is 2 cm × 2 cm. We have achieved the scanning range of about 3 mm within the scanning mirror's vibration angle of approximately 10°.

Abstract: Example 11 (136 words)

We have presented a simple method to prepare tungsten nanotips. The method was based on a "drop-off" method, in which the tip was continuously and slowly drawn up from the electrolyte during etching. We then observed the field emission current with respect to the applied voltage from the fabricated tip with and without the irradiation of 20 fs laser pulses at a wavelength of 800 nm. The tips showed good field emission properties as revealed by the current–voltage characteristics and analyzed based on the Fowler–Nordheim theory. The turn-on field was 5 V/μm to achieve a current density of 500 A/m^2, whilst the local electric field strength at the tip was equal to 2 V/nm. Together with their ease of preparation, these tips were shown to have great potential in the area of electron field-emitting devices.

Abstract: Example 12 (76 words)

The fluorescence temporal profile has been studied at the symmetrical point of the target in tissue phantom by a time-resolved fluorescence diffuse optical tomography method. Indocyanine green served as the fluorescence target. The results showed that the geometrical symmetry of the fluorescence peak intensity ratio was broken due to target position. By using this parameter, an unknown target location could be identified for fluorescence targeting and further reconstruction of the fluorescence image in a tissue model.

Abstract: Example 13 (117 words)

We have prepared zinc oxide nanoparticles by using the pulsed laser ablation approach (Nd^{3+}:YAG laser at a central wavelength of 1064 nm with the pulse energy of 100 milli-joules) from the zinc metal plate immersed in a solution of sodium dodecyl sulfate at the ambient temperature. The laser fluence dependence of absorbance of the produced nanoparticles shows a single and sharp peak around 375 nm, indicating that the nanoparticles have a narrow size (45–60 nm) with almost spherical shape. Next, zinc oxide nanoparticles with different concentrations can hold promise in the phototoxic effect. The photocatalysis activity of the zinc oxide nanoparticles can considerably enhance the injury of a cancer cell mediated by the reactive oxygen species.

Abstract: Example 14 (104 words)

We have studied the influence of shape and chemical content of metallic nanostructures (gold–palladium core–shell nanorods, gold nanobipyramids, and single-crystalline porous palladium nanocrystals) on the luminescence of Rhodamine 6G and Coumarin 153 dyes. The changes found in the emission intensity detected by the optical spectroscopy are attributed to different proportions of the dyes and the metallic nanostructures. The enhancement of the fluorescence observed by us for the cases of Rhodamine 6G with gold–palladium core–shell nanorods and Rhodamine 6G with gold nanobipyramids suggests their promising plasmonic properties. This may be regarded as a preliminary step toward development of the fluorescent probes possessing high photo-sensitivity.

Abstract: Example 15 (119 words)

We focus on the plasmonic properties of palladium-coated gold nanorods. Two characteristic plasmon bands of those nanorods have been detected in the optical absorption by using the optical spectroscopy. One of them, which is located at about 525 nm, is associated with electron oscillations along the transverse direction. This band does not depend on the palladium-shell thickness and the dielectric susceptibility of the surrounding medium. The other band, at about 800–900 nm, is associated with electron oscillations along the longitudinal direction. It exhibits a clear shift when the palladium-shell thickness and the dielectric surrounding are changed. Our results point to a novel possible way for tuning photo-catalytic ability of the nanorods and their possible use in biological/chemical sensors.

3.2   Conclusion

Conclusion is the last paragraph in a paper. The conclusion is like the final chord in a song. It makes the listener realize that the piece is completed. You want them to understand the issues you have raised in your paper, synthesize your thoughts, and show the importance of your ideas (your findings). You then become a reliable author and create a lasting impression for them. They are more likely to read your paper in the future. They may also have learned something, and what you have reported might have changed their opinion. The conclusion is, in some ways, like the introduction section plus the abstract section. You can restate the main points of evidence in these sections for the readers. Finally, you can report conflict of interests (financial, personal assistance) in the acknowledgment section after the conclusion section.

When you write the conclusion, you should avoid some tactics:

1. You should not make sentimental, emotional appeals that are out of character with the rest of the analytical paper.
2. You should exclude the evidence (i.e., quotations, statistics, etc.) that should be in the body of the paper.

3.2.1 Some Examples of Conclusion

Conclusion: Example 1 (90 words)

Shadow deposition has been demonstrated as a versatile technique to prepare various kinds of Cu nanodots and Cu nanowires with a desired structure. Thinner nanowires show stronger anisotropic optical absorption than the thicker ones. Absorption maxima for the polarization parallel to the wire axes occur at the lower photon energy than that for the polarization perpendicular to the wire axes. The red-shift of the absorption maximum is observed when the widths of the nanowires are increased. These experimental results are not consistent with the theoretical prediction by the Maxwell–Garnett theory.

Conclusion: Example 2 (85 words)

For arrays of Cu nanowires, we have shown that plasmon maxima for the incident electric field polarized parallel to the long wire axes appear at lower photon energy than the maxima for perpendicular polarization. By increasing the nanowire widths, plasmon maxima around 2.3 eV are shifted to lower photon energy. This experimental dependence of the energy positions of spectral absorption maxima on the light polarization and the wire width cannot be explained completely by the Maxwell–Garnett model but can be explained well by FDTD calculations.

Conclusion: Example 3 (113 words)

We have measured azimuthal angle dependence of the second harmonic intensity from Cu nanowires in air at the fundamental photon energy of 1.17 eV. We see two main lobes in p-in/p-out, s-in/p-out, and s-in/s-out polarization configurations. The second harmonic intensity is as low as the noise level for p-in/s-out polarization configuration. Using a phenomenological model, we have found that the contributions of χ^2_{222}, χ^2_{311}, and χ^2_{323} elements dominate the second harmonic signal for Cu nanowires. The χ^2_{222} element corresponds to the consequence of the breaking of symmetry in the wire structures. The second harmonic response from the electromagnetic field gradient normal to the substrate plane is suggested to enhance the SHG from the χ^2_{311} and χ^2_{323} elements.

Conclusion: Example 4 (76 words)

We have measured the photon energy dependence of the second harmonic intensity from Cu nanowires. The main peaks of the spectra of Cu nanowires are found at two-photon energy of 4.4 eV for the p-in/p-out and s-in/p-out polarization combinations. Theoretical calculation by a local field model indicates that at this photon energy the observed second harmonic response is induced by a resonant coupling between the fundamental field and the surface plasmon in the wires.

Conclusion: Example 5 (62 words)

We have tested the concept of total-light algorithm with tissue phantoms. The reconstruction image by total light can eliminate the absorption of the fluorophore and then the target can be visualized by the subtraction of the images between the absorption images obtained by the excitation and total light. This method is potentially useful to visualize the fluorophore absorption in the inhomogeneous tissue.

Conclusion: Example 6 (53 words)

Our first trial for the image reconstruction with a fluorescence target implanted into a rat abdominal cavity in vivo has been successfully demonstrated. The total-light approach will eliminate the problem of the inhomogeneity of the tissue and quantify the fluorescence target. Therefore, this approach is another option for fluorescence diffuse optical tomography.

Conclusion: Example 7 (140 words)

We have proposed and constructed an optical delay line for rapid scanning low-coherence reflectometer. In the optical delay line, a retro-reflector and a reflection mirror were integrated in an aluminum jig. A galvanometer was used to drive the scanning mirror as a scanning component, which performed a scanning rate of 400 Hz in our system. The size of the optical delay line is within 2 cm × 2 cm. As the vibration angle of the scanning mirror is 9.6°, the achieved scanning range of the low-coherence reflectometer can be larger than 3 mm. The optical delay line possesses the advantages that it is compact, easy to fabricate, and can avoid walk-off problem during scanning. The low-coherence reflectometer with the proposed optical delay line can perform a rapid scanning imaging and two-dimensional image of a stack of coverslips was demonstrated.

Conclusion: Example 8 (117 words)

Polycrystalline AlN films on the quartz substrates were deposited using an aluminum target. The depositions performed in our home-made RF magnetron sputtering system under various nitrogen concentrations. The X-ray diffraction patterns showed the peaks corresponding to (002), (102), and (103) reflections. The intensity of the (002) diffraction peak was the most intense at nitrogen concentration of 40%, indicating our improvement of the film's crystallinity. The deposited films also showed the high transparency in visible wavelength. Furthermore, the refractive index and optical band gap energy of the films as a function of nitrogen concentration were determined. It was found that refractive index weakly depended on nitrogen concentration, whilst the optical band gap energy strongly depended on nitrogen concentration.

Conclusion: Example 9 (170 words)

Aluminum nitride films on unheated quartz substrates were deposited by reactive radio frequency magnetron sputtering process under argon/nitrogen mixture atmosphere. The effect of nitrogen concentration on the bonding formation of aluminum nitride, structure and morphology of the films was studied. The Fourier-transform infrared and Raman spectra of the films confirmed the absorption band corresponding to E1 (TO), A1 (TO), and E2 (high) vibration mode of the infrared active bond of aluminum nitride. The crystallographic orientation of the films depending on the nitrogen concentration was optimized under nitrogen concentration of 40%. From the cross-sectional field emission scanning electron microscope image of the film with this optimized film condition, the columnar structure was visibly observed. The dense columnar grains were uniformly seen on the film surfaces fabricated under all nitrogen concentration conditions, except for nitrogen concentration of 20%. The bulk resistivity and piezoelectric property were investigated through the metal–insulator–metal structures embedded by the aluminum electrodes. The resistivity was in a range of 10^{14}–10^{15} Ωcm. The effective piezoelectric coefficient d_{33} was 11.03 pm/V.

Conclusion: Example 10 (103 words)

We have constructed the optical coherence tomography system with the optical delay line consisting of the reflective components for large scanning range with high frequency. In the delay line, the retro-reflector together with the inclined reflection mirror is assembled onto the aluminum jig. The dimension of our delay line is 2 cm × 2 cm. When the vibration angle of the scanning mirror is about 10° at a scanning rate of 400 Hz, we have achieved the imaging depth of the optical coherence tomography system at approximately 3 mm. Our optical delay line has advantages that it is very compact and easy to fabricate.

Conclusion: Example 11 (91 words)

We have reported a study of field emission from the tungsten nanotips synthesized by electrochemical process. The experimental results showed that the field emission behaviors of the tips were in agreement with the Fowler–Nordheim theory. The turn-on field at a current density of 500 A/m^2 was about 5 $V/\mu m$, while the local electric field strength at the tip was equal to 2 V/nm. The excellent field emission performances indicated that the tungsten nanotips grown by the present approach were a good candidate for application in the area of electron field-emitting devices.

Conclusion: Example 12 (83 words)

Time-dependent measurement of fluorescent light propagation from indocyanine green target in tissue phantom was carried out in order to investigate the fluorescence temporal profile at the symmetrical point of target. The peak intensity ratio of the fluorescence temporal profile was then analyzed. It was shown that the geometrical symmetry of this parameter was broken because of target position. Using this asymmetry feature, an unknown target position was identified. Furthermore, this analysis would suggest better excitation detection geometry for fluorescence diffuse optical tomography research.

Conclusion: Example 13 (174 words)

We have produced zinc oxide nanoparticles in a sodium dodecyl sulfate solution as a surfactant using the laser ablation technique at room temperature. The produced zinc oxide nanoparticles have spherical shapes. The typical diameters of the zinc oxide nanoparticles are about 45–60 nm. The absorbance of the zinc oxide nanoparticles increases with the laser fluence. The single sharp peak of the absorbance centered at 375 nm reveals that the nanoparticles have a narrow size with nearly spherical shape. This work also reports our findings on a potential application of the zinc oxide nanoparticles in the biomedical application in cancer therapy. The experimental result demonstrates that ultraviolet irradiation can induce the growth suppression of cancer cells by photocatalysis activity of the nano-sized zinc oxide. This ultraviolet irradiation suppresses the viability of the MCF-7 cells incubated with different concentrations of the zinc oxide nanoparticles, suggesting a concentration-dependent effect. Furthermore, the zinc oxide nanoparticles can induce levels of generation of oxidants, such as the reactive oxygen species, proposed as the common mediators for the phototoxic effect.

Conclusion: Example 14 (181 words)

We have described two observations regarding the optical properties of the gold–palladium core–shell nanorods. The nanorods reveal a unique optical response, with the two characteristic plasmon resonance bands present in the optical absorption. Unlike the resonant peak from the transverse plasmon band, the peak linked with the longitudinal plasmon band excitation shows a remarkable shift when one changes the palladium-shell thickness and the dielectric medium in a controllable manner. A blue shift of the localized longitudinal surface plasmon resonance in the UV–Vis–NIR absorption spectrum is observed if more palladium atoms are added in order to form a shell on the gold core. This allows one to alter the plasmon resonance of the palladium-coated gold nanorods. On the other hand, since high enough refractive indices of the organic solvents can screen the incident electromagnetic field, the localized longitudinal surface plasmon resonance becomes red-shifted with increasing refractive index of the surrounding medium. Due to excellent reproduction of the absorption response and the intense plasmon peaks, the palladium-coated gold nanorods can serve as a superior candidate for chemical/biological sensing and photocatalysis.

Conclusion: Example 15 (101 words)

In the present work, we have investigated potential enhancement of the fluorescence emission of the standard dyes Rhodamine 6G and Coumarin153 doped with the metallic nanostructures of three different shapes. These are gold–palladium core–shell nanorods, gold nanobipyramids, and single-crystalline porous palladium nanocrystals. We have obtained the highest (more than threefold) fluorescence enhancement on the system of Rhodamine 6G mixed with the gold–palladium core–shell nanorods. On the contrary, the smallest fluorescence changes have been observed for Coumarin153 that involves the porous single-crystalline palladium nanocrystals.

3.3 Grammar

In the paper, you should use the correct tense in constructing sentences. You should use simple past for specific methods, simple present for facts, and present perfect for general findings or many studies. Here are some examples of grammar references.

Present Simple

(i) Regular verbs add -s or -es in the present simple positive.

I/You/We/They	*He/She/It*
observe	observes
display	displays
exhibit	exhibits
observe	observes
come	comes
buy	buys
see	sees
go	goes

The form is the same for all persons.

I/You/We/They	observe the surface plasmon resonance.
He/She/It	observes the surface plasmon resonance.

(ii) Present simple negative

I/You/We/They	*He/She/It*
do not	does not

The form is the same for all persons.

I/You/We/They	do not observe the surface plasmon resonance.
He/She/It	does not observe the surface plasmon resonance.

Present Perfect

(i) Regular verbs add -d or -ed in the present perfect positive.

I/You/We/They	*He/She/It*
observe	observed
display	displayed
exhibit	exhibited
observe	observed

(ii) Many common verbs are irregular.

Present	*Past participle*
come	come
buy	bought
see	seen
go	gone

(iii) Irregular verbs do not add -d or -ed in the present perfect.

I/You/We/They/He/She/It

come

bought

seen

gone

(iv) The form is the same for all persons.

I/You/We/They	have observed the surface plasmon resonance.
He/She/It	has observed the surface plasmon resonance.

(v) Present perfect negative

I/You/We/They	*He/She/It*
have not	has not

The form is the same for all persons.

I/You/We/They	have not observed the surface plasmon resonance.

| *He/She/It* | has not observed the surface plasmon resonance. |

Past Simple

(i) Regular verbs add -d or -ed in the past simple positive.

Present	*Past*
convey	conveyed
display	displayed
exhibit	exhibited
observe	observed

(ii) Many common verbs are irregular.

Present	*Past*
come	came
buy	bought
see	saw
go	went

(iii) The form is the same for all persons.

| *I/You/We/They* | observed the surface plasmon resonance. |
| *He/She/It* | observed the surface plasmon resonance. |

(iv) Past simple negative

| *Present* | *Past* |
| do/does not | did not |

The form is the same for all persons.

| *I/You/We/They* | did not observe the surface plasmon resonance. |
| *He/She/It* | did not observe the surface plasmon resonance. |

3.4 Problems

Pb.1. Select the abstract of your paper and write the conclusion.

Pb.2. Select the conclusion of your paper and write the abstract.

Pb.3. Discuss the content of the abstract.

Pb.4. Discuss whether the following abstracts (A –I) are suitable or not:

A. We demonstrate that the local field near Cu nanowires enhanced by the plasmon excitation gives rise to second harmonic response when the excitation optical field is polarized perpendicular to the wire axes.

B. We have studied rotational anisotropy in the second harmonic (SH) intensity from Cu nanowires. The analysis of our experimental result by finite-difference time-domain (FDTD) method has indicated that the SH response from Cu nanowires is enhanced by the coupling of the electric field components along the surface normal to the local plasmons and the electrostatic lightning-rod effect.

C. We have obtained the first confocal microscopic sum frequency (SF) intensity images, using ZnS polycrystals as a sample. We have found different contrasts in the SF intensity images when the incident beams were focused at different depths in the ZnS sample.

D. We have demonstrated confocal sum frequency (SF) microscopy. The SF intensity images were obtained when the incident beams were focused at different depths in the ZnS sample from 0 to 10 μm.

E. We have fabricated Au, Cu, and Pt nanowires by using a shadow deposition technique. We then investigated optical properties of these nanowires by studying the second harmonic generation (SHG) spectra. With experimental and theoretical studies, we have discovered that the surface plasmon resonance was served as an origin of the enhancement of the SH response from the metallic nanowires.

F. The strongly geometrical shape, high aspect ratios, and nanoscale cross-section of nanowires are expected to affect optical properties through confinement effects. Herein we have investigated optical properties of Cu nanowires with studies of second harmonic generation (SHG) spectra. These optical properties of Cu nanowires will be one of the most important issues when considering the types of materials used in current applications and development of new applications.

G. We studied optical properties of copper nanowires fabricated by a shadow deposition method. We found that the optical absorption spectra of thinner nanowires exhibited stronger anisotropic absorption than thicker nanowires. Absorption maxima are located at lower photon energy when incident field is parallel to the wire axes than they are perpendicular to each other. They also shift to lower energy when the widths of nanowires are increased.

H. We have performed a target size dependence of fluorescence temporal profile from an indocyanine green (ICG) target filled in a phantom by fluorescence diffuse optical tomography (FDOT) method. From the results of experiment and statistical analysis, we have found that the geometry effect of target size was one of the key parameters in temporal profile determining the resolution of the optical reconstruction images.

I. The second-order responses depend on the structural symmetry of the metal nanostructures, which can give rise to interesting polarization dependences in the responses. We also show that the sensitivity of second-order processes to the symmetry provides important information about plasmonic effects in the nonlinear properties of the structures.

Pb.5. Discuss whether the following conclusions (A–I) are suitable or not:

A. We have studied the azimuthal angle dependence of SH intensity from Cu nanowires. We have found that χ^2_{222} is dominant due to the broken symmetry of the surface structures, while χ^2_{311} and χ^2_{323} strongly depend on the dipole interaction between nanowires by local field enhancement associated with plasmon excitation

B. We combined the results of SH intensity pattern analysis and FDTD simulations to gain insight into the field enhancement in Cu nanowires. We have found that the enhancements of the χ^2_{311} and χ^2_{313} elements were induced by the lightning-rod effect and the local plasmons in nanowires, respectively.

C. We have constructed a confocal sum frequency (SF) microscope for the first time and have obtained the images of SF intensity from ZnS polycrystals. The confocal SF microscopy allowed acquisition of images of SF intensity of ZnS planes of different depths.

D. We have constructed a confocal sum frequency (SF) microscope and have obtained cross-sectional SF intensity images from ZnS polycrystals at several different depths.

E. We have studied optical properties of self-organized arrays of metallic polycrystalline nanowires: Au, Cu, and Pt fabricated by the shadow deposition technique. We have suggested that the enhancement of local field by plasmon excitation in the nanowires was a candidate origin of the observed SH intensity.

F. We studied nonlinear optical properties of Cu nanowires prepared by the shadow deposition method. The origins of the observed SH intensity from nanowires were the enhancement of the local field by the plasmon excitation in the nanowires.

G. Shadow deposition has been demonstrated for fabrication of Cu nanodots and nanowires. Thinner nanowires show stronger anisotropic optical absorption than the thicker ones. Absorption maxima for polarization parallel to the wire axes occur at the lower photon energy than that for polarization perpendicular to the wire axes. The red-shift of the absorption maximum is observed when the widths of the nanowires are increased. These experimental results are not consistent with theoretical prediction by the Maxwell–Garnett theory.

H. The fluorescence temporal profile as a function of geometrical configuration was investigated by FDOT method, and it was sensitive to variation of the ICG target size inserted within a cylindrical POM phantom. This was a prerequisite information to acquire the accuracy of tomographic images of the fluorescent regions in tissue samples. Using an algorithm based on the photon diffusion equations, the optical images of the phantoms will be able to be reconstructed.

I. From the analysis of the second harmonic intensity patterns by the phenomenological model and the finite-difference time-domain method, we have concluded that second harmonic generation from Cu (Pt) nanowires was found to be different from those of Au nanowires. This was due to the weaker suppression of the incident field by the depolarization field in Cu (Pt) nanowires and the stronger anisotropy in the nonlinearity of Cu (Pt) nanowires due to thinner wires than Au. The electric field component along the surface normal enhanced by the plasmon-resonant coupling in the wires was observed for Cu (Pt) nanowires, but it was not for Au nanowires.

Chapter 4

Introduction to Report

After you select the topic of your research from the defined problem (basic and/or applied research), you have to search for all relevant literature related to your topic. This will get the reader's attention and interest. This chapter will show how to carry out literature survey.

4.1   Literature Survey and Developing a Conceptual Framework

You first build your research framework of relevant and current studies (summary of previous research) that you may get from journals, textbooks, and the Internet. Then you use the framework to write your literature review cogently, which will lead your argument to research objectives, research questions, or hypothesis. The framework can be conceptual. You must draw a simple diagram showing how the variables in your study are related to the theories (see below).

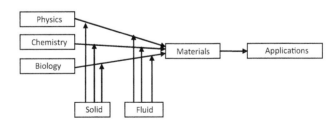

Research Methodologies for Beginners
Kitsakorn Locharoenrat
Copyright © 2017 Pan Stanford Publishing Pte. Ltd.
ISBN 978-981-4745-39-0 (Hardcover), 978-1-315-36456-8 (eBook)
www.panstanford.com

4.1.1   Examples of Conceptual Frameworks

Conceptual Framework: Example 1

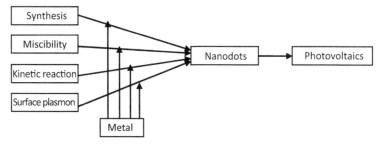

Conceptual Framework: Example 2

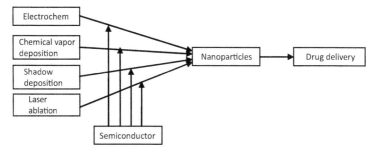

Conceptual Framework: Example 3

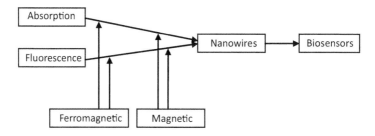

4.2   Introduction of the Problem

The statement of the problem could be a description of the current situation or a gap in scientific knowledge. It could be a description of a current controversy or contradiction in findings in the area of interest. It also could be an explanation of how the current findings or beliefs are no longer true in the light of recent developments. You can show here the reasons why the study should be undertaken, how urgent or important it is, and what contributions your research will make to your area of study. You can indicate the potential practical or policy implications of your research. You can show intellectual merit: How your research will make an original contribution? How it will fill gaps in the existing research? How it will extend understanding of particular topics? After that, you write the research objectives, research questions, or hypothesis. In this section, you use the statement: "Therefore, this study intends to …" Finally, you may briefly describe the research methodology or outline of your present work and the outcome of the study.

4.2.1 Examples of Introduction

Introduction: Example 1

Interests in nanotechnology have inspired a lot of researches on the metal nanowires and opened up a new prospect of applications in future electronic and optical devices [1,2]. Cu is of particular interest because it can be suitable for potential application such as in interconnection in electronic circuits. The developments of the fabrication approaches to produce Cu nanowires have been reported in literatures [3–5]. For the optical application requiring nanowire arrays of large areas with a high throughput, shadow deposition technique by templating against relief structures on the (110) surfaces of NaCl crystals has been proposed [6].

Highlight the recent trends

So far, many optical studies on the metal nanowires have been performed. Sandrock et al. have disclosed the local-field enhancement of noncentrosymmetric Au nanostructures [7], Schider et al. have shown the excitation of dipolar plasmon mode of Au and Ag [2], and Kitahara et al. have observed optical nonlinearity from Au nanowire arrays [8]. Despite a number of researches on the bulk properties of Cu films [9,10], no literature can be found on the optical properties of Cu nanowire arrays and hence there is a need to focus on such nanowire materials.

Need for study

Herein this study intends to report the preparation and characterization of Cu nanowires and nanodots on the NaCl(110) faceted surfaces prepared by the shadow deposition process. Absorption spectra of normal incidence are measured as a function of the sample rotation angle. The comparison between the experimental findings and the calculation by Maxwell–Garnett model are discussed.

Objective of research

Introduction: Example 2

Interaction of light with metal nanowires has attracted attention through its applications to nonlinear optics [1], surface-enhanced Raman scattering [2], and plasmonics [3]. These nanowire systems strongly absorb incident light at frequencies of surface plasmon resonances. Although the optical properties of metal spheres are described by Mie's theory in 1908 [4], the relationship between the wire geometries and their linear optical properties has not been fully investigated. Hence, the production of metal nano-wires with specific size, geometry, and distribution is important to their theoretical research and technological applications utilizing surface plasmon resonance. Compared with physical methods such as electron beam lithography [5], shadow deposition is considered as a much more efficient method allowing the deposition of large areas of anisotropic nanostructures and presenting high fabrication throughput.

Highlight the recent trends

Need for study

Cu nanowires are particularly interesting to study because the confinement of the charge carriers is prominent to such an extent that the electronic states change remarkably [6]. Our previous report has shown that on facetted NaCl(110) surfaces well-organized arrays of Cu nanowires grow and their absorption spectra as a function of photon energy show strong plasmon resonance spectra [7]. However, we found that the detailed experimental dependence of the peak energy positions of the absorption maxima on the polarization is different from the theoretical prediction by Maxwell–Garnett model. In this effort, we intend to discuss the observed nanowire's plasmon resonance through a calculation by finite-difference time-domain (FDTD) method. In such an analysis, we may be able to consider a more realistic scenario in the dependence of their absorption spectra on the wire sizes.

Objective of research

Introduction: Example 3

Recent advances in nanofabrication techniques are ready for generating controlled systems of metals. Especially, much effort has been performed to investigate their nonlinear optical properties by second harmonic (SH) spectroscopy [1,2]. A great advantage of these studies over the linear optical techniques is its intrinsic surface sensitivity due to the fact that the SH signal originates from a thin interfacial region. This field is expected to transit soon from fundamental understanding of its principles to useful applications [3].

Information on nonlinear optical properties can be obtained by rotating the crystalline metal around its surface normal while measuring the modulation of the SH intensity. For instance, Tom and Aumiller [4] have shown that the rotational anisotropy in SH intensity of copper films is primarily attributed to a resonance between the harmonic field and the d bands of copper. On the other hand, when one considers metal nanowires, the optical patterns can be modulated by the effect of the wire shapes: The fields near a surface with a higher curvature may be intrinsically higher. The excitation of propagating or localized surface-plasmon waves can occur if the metal wires act as a grating coupler. The influence of retardation on the surface charge excitations is important for larger wires [2]. All of these factors motivate the need for further studies toward the elucidation of the microscopic origin of the nonlinearity of metallic nanostructures.

In this study we intend to perform a rotational anisotropy experiment of the SH intensity from arrays of copper nanowires. According to an experimental observation and a theoretical work, we have concluded to have observed second harmonic generation enhanced by the electromagnetic field gradient normal to the substrate plane.

Introduction: Example 4

Nonlinear optical techniques have been frequently used in recent years to study surface phenomena. In particular, second harmonic generation (SHG) has been established as a very powerful spectroscopy to study a wide range of physical natures at the surface and interface structures of solid state materials [1]. Within the electric dipole approximation SHG occurs only if the medium lacks inversion symmetry. Therefore, the nonlinear susceptibility vanishes in the bulk of centrosymmetric media but has a finite value at the surface where the symmetry is broken.

Some of the SHG studies are concerned with noble metals [2–4]. Namely, Lüpke et al. have shown that the SH resonance from the bulk Cu was attributed to a coupling between the SH photon and the electronic transition [4]. However, the metallic nanostructures generally have different structural symmetries from the bulk [5]. The high-density gradient and strong delocalization of the electrons at the surface may contribute to the SHG enhancement. The plasmon contribution is probably operative for enhancing the second harmonic nonlinearity. Consequently, it is strongly expected that the study of the electronic spectra of these nanostructures may clarify the mechanism of the nonlinear optical response of metallic nanomaterials.

Need for study

In this paper, we intend to measure the spectral dependence of the SH signal from Cu nanowires as a function of the SH photon energies from 2.4 to 4.6 eV. The experimental results are compared with a local field model for understanding of the electronic excitation on the surface of these metallic nanostructures.

Objective of research

Introduction: Example 5

Fluorescence imaging technique is a key technique for the molecular imaging in biological system and the application is extending to living systems from small animals to human tissues [1,2]. The main problem of the fluorescence imaging in tissue is the strong distortion by scattering and absorption; the diffusive light transport eliminates the detail spatial information. In case of the small animal imaging, such as mice, there are already some commercial products, which construct image of fluorescent targets. The technology is rather simple but it is known to be practically useful. This is because the scattering problem in the fluorescence imaging of the mice size systems is not big. However, it is really difficult to extend to larger systems. It is obvious that the imaging should take the diffusive and inhomogeneous properties into account to recover the lost spatial

Highlight the recent trends

Need for study

information and the distortion. This problem is the main subject of fluorescence diffuse optical tomography (FDOT).

In the strong scattering system, the fluorescence photon propagation can be described by a joint partial differential equation, which consists of two differential equations of the excitation light and of the fluorescence photon generated by the excitation along the excitation light path. The inverse problem for them is actually very hard task. Recently, Marjono et al. have proposed a new algorithm, total-light algorithm, to simplify the equation [3]. The basic idea of this algorithm is following: since the energy loss of the excitation light by fluorophores are converted to the fluorescence emission, this loss can be estimated by the observed emission energy if the optical properties of the background are same at the excitation and emission wavelengths and the quantum efficiency is known. Finally, the joint equation can be reduced to two independent diffusion equations; the problem is reduced to a diffuse optical tomography (DOT) problem.

Since the algorithm is a variation of diffuse optical tomography, the sensitivity is also limited by the original DOT. However, this method can separate the absorption of the fluorophores from the total absorption, which can be obtained by the normal DOT, resulting that this algorithm can quantify the fluorophores.

As mentioned above, the quantum efficiency and the fluorescence lifetime of dye should be known. Further, the optical properties should be same at the excitation and the emission wavelengths, except the fluorophore absorption at the excitation wavelength. In the real samples, these assumptions do not somewhat hold and the violation of them may cause the degradation of the images. In this study, we intend to investigate the applicability of our algorithm. Further, we have focused on how much geometrical information the fluorescence carries to improve the algorithm. In this paper, we will show the results of the temporal profile changes with single target phantoms. Research objective
We will confirm the algorithm assumption and then discuss the reconstructed images.

Introduction: Example 6

Fluorescence imaging is actively investigated as a key technique of the molecular imaging in living systems and has been applied from small animals to human tissues with the aim of the detection, diagnosis, and/ or monitoring of cancer and other disorders.[1,2] One of the challenges in the quantification of the fluorescence images in tissue is the strong scattering and absorption, which is the most difficult problem to overcome. The details of the fluorescence distribution are lost by the diffusive property of the photon transport in tissue and eventually the simple projection approach of the fluorescence image is not quantitative. It is still useful for the animals like mice, but difficult to extend to more larger system like human tissue. It is obvious that the technique, which will use for the imaging of the large systems, need to account the diffusive and inhomogeneous properties of tissues to recover the lost spatial information and the distortion. That technique is called fluorescence diffuse optical tomography (FDOT) or fluorescence molecular tomography (FMT).

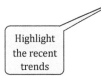

In this study, we intend to focus on FDOT technique using the total-light approach.[3] The so-called total-light approach can be categorized in the FDOT, but this approach actually reconstructs the distribution of the absorption of the fluorophores in tissue by using the conventional DOT. In contrast to the conventional DOT, this approach can separate the absorption of fluorophores from the tissue. This will be an advantage of the total-light approach because the tissue absorption is usually very complex due to its inhomogeneous nature.

The reliability of the present technique has been confirmed with the known fluorescence target in tissue phantom. Total-light approach is applied to a fluorescence target implanted into the abdominal cavity of rat in vivo. At first, the fluorescence decay profiles are discussed, and then the reconstructed images are presented.

Introduction: Example 7

An optical delay line is an optical component which can provide phase variation of an optical signal via variable optical path length. In practice, the optical path length can be changed discretely or continuously. Imaging techniques such as white-light interferometry [1] and phase-shifting interferometry [2] usually need optical components for discrete phase shift. On the other hand, low-coherence reflectometry [3] and phase-resolved interferometry [4] usually need a continuous scanning optical delay line to provide an axial scanning.

Traditionally, the optical delay can be achieved with a piezoelectric transducer (PZT), a stepper motor stage or a linearly motorized translation stage. The piezoelectric transducer can perform a rapid scanning due to its high vibrating frequency. However, the traveling range of a piezoelectric transducer is usually limited in the scale of micrometer. On the other hand, the stepper motor stage and linearly motorized translation stage can perform long-range scanning, but the translation speeds of such stages are usually limited in the order of centimeter per second. In many applications, an optical delay line which can perform a rapid scanning in the millimeter range is valuable. Furthermore, the optical delay line has to be stable, compact and easy for maintenance. Many different designs of optical delay lines were proposed for large scanning range with high frequency. A diffraction grating-based rapid-scanning optical delay line (RSODL) was proposed for the axial scanning of time-domain optical coherence tomography [5,6]. In the typical configuration of an RSODL, a diffraction grating and a scanning mirror are arranged on two focal planes of a lens, respectively. By use of such an arrangement, the group-delay and phase-delay of the optical signal can be controlled independently. However, even if an achromatic lens is introduced to reduce the chromatic aberration, the offset of the ray from the pivot point of the scanning mirror will cause a walk-off effect. In other words, the direction of the beam returned from the diffraction grating deviates from the direction of the incident beam and leads to stray intensity modulation during scanning [7].

Recently, various configurations for the optical delay lines were developed to enhance the scanning rate and scanning range such as using a rotating cube [8], a polygonal scanner [9–11], a multiple-pass cavity delay line [12], or a combination of curved mirror and scanning mirror [13,14]. Several groups used a rotary array of mirrors [15,16], prisms [17,18], or corner reflectors [19] to enhance the scanning rate.

Although most of these techniques can provide high-frequency and large-range scanning, some configurations are difficult to fabricate or difficult to construct in small size. Furthermore, the walk-off problem usually causes variation in intensity of optical signal during scanning which may result in measurement errors when the intensity of light is under investigation.

In this study, we intend to demonstrate an all reflective optical delay line which is compact, easy to fabricate, and can avoid the intensity loss during scanning. The scanning range of 2.9 mm is achievable when the vibration angle of the scanning mirror is 9.6°. The scanning range can be further increased when the pivot of the scanning mirror deviates from the incident beam. The optical delay line provided a scanning rate of 400 Hz and is demonstrated with a low-coherence reflectometer. A higher scanning rate can be achieved when a galvoscanner with higher scanning rate is applied.

Objective for study

Introduction: Example 8

Aluminum nitride (AlN) thin films are well-investigated materials for potential applications. This is because their outstanding physical properties, such as high melting point, wide energy band gap energy, high static dielectric constant, and low thermal expansion coefficient [1,2]. Due to these specific properties, AlN thin films have drawn considerable interest and large number of publications have been devoted on their film preparations under various substrates.

Highlight the recent trends

So far, different methods have been carried out for such films, such as, halide vapor phase epitaxy (HVPE) [3], metal-organic chemical vapor deposition (MOCVD) [4], pulsed laser deposition (PLD) [5], molecular beam epitaxy (MBE) [6], single ion beam sputtering (SIBS) [7], and double ion beam sputtering (DIBS) [8]. However, they have drawback because the specimens have been prepared under high temperature conditions. This might lead to degradation of the substrates and the films during deposition. Therefore, low temperature preparations of the AlN films are more attractive and very important. In this contribution, we then introduce radio frequency reactive magnetron sputtering to fabricate the AlN films under low temperature conditions.

Need for study

This study intends to investigate nitrogen concentrations (CN) on structural and optical properties of polycrystalline AlN films on the quartz substrates using RF reactive magnetron sputtering.

Objective for study

Introduction: Example 9

Aluminum nitride (AlN) films have received a steadily growing interest as a result of their fascinating properties such as high chemical stability, high thermal conductivity, wide direct band gap, high electrical resistivity, high acoustic velocity, and high piezoelectric coefficients. Thus, they have been used as insulating layers, optical sensors, surface acoustic wave devices, and in MEMs applications [1–5]. Many fabrication techniques, such as sputtering, chemical vapor deposition, pulsed laser deposition, and molecular beam epitaxy, have been performed to prepare the AlN films on the different substrates; however, the purposed deposition temperatures were very high resulting in degradation of the substrates and the thin films. Hence, fabrication of the AlN films by the radio frequency reactive magnetron sputtering technology under a low temperature is presented in this contribution.

This study intends to present the deposition of the AlN thin films on the quartz substrates by radio frequency reactive magnetron sputtering technology under a room temperature. The films properties will be then investigated by X-ray diffraction, FTIR and Raman spectroscopy, atomic force microscopy (AFM), and field emission scanning electron microscopy (FE-SEM). The bulk resistivity and piezoelectric properties were also studied via the metal–insulator–metal structures.

Introduction: Example 10

Many optical imaging techniques, for instance phase-shifting interferometry [1–6] and low-coherence reflectometry [7–12] require the effective optical delay lines offering a phase variation of optical signal through discretely or continuously optical path length. Originally, many designs of the delay lines have been proposed either for large scanning range or for high frequency. For example, the piezoelectric transducer performs a rapid scanning, but the traveling range is restricted in micrometer scale. The stepper motor stage and linearly motorized translation stage performs long-range scanning; however, the translation speed is limited in the scale of centimeter per second. So far, different designs of the delay lines have been developed to increase both scanning rate and scanning range [13–23]; however, most of these configurations are not easy to fabricate into small components.

In this article, we intend to construct the compact optical delay line for the axial scanning of time-domain optical coherence tomography that is very easy for fabrication. Another benefit of our delay line is that the scanning range is large in the millimeter range when the vibration angle of the scanning mirror is about 10° with the high scanning rate of 400 Hz.

Objective for study

Introduction: Example 11

Field emission involves a quantum mechanical tunneling process under applied field. It has diverse technological applications in the area of electron field–emitting devices. In field emission, a strong static field is able to bend the potential at the surface of a nanoscale tip so that electrons are emitted by quantum tunneling. The apex of the tip also must be much smaller than the wavelength of the optical radiation, so the time-dependent electric field of the radiation is superimposed on the applied static field to modulate the current of the emitted electrons. For field emission from a nanotip configuration, the tip radius was a key factor for its field emission behavior [1]. In order to obtain excellent field emission performances, extensive effort is needed to synthesize the sharp tips with a nanometre-scale radius of curvature.

Highlight the recent trends

Need for further study

Huge number of works on the fabrication of the sharp tips have been performed and several techniques have been developed, such as grinding [2], mechanical pulling [3], chemical or electrochemical etching [4–6], and ion milling [7]. Among these methods, the etching technique is most widely used as a fast, cheap, convenient, and reliable method. One of the common and reliable methods is so called a "drop-off" method, in which etching occurs at the air–electrolyte interface, causing the portion of wire immersed in the solution to "drop off" when its weight exceeds the tensile strength of the etched wire metal neck [8]. This method is also suitable for different kinds of metals and will be used in our experiment.

This paper intends to investigate the field emission properties of tungsten nanotips produced by electrochemical etching with and without the irradiation of intense femtosecond light pulses in near infrared. We will show that the tips can start to emit current by applying the voltages above 5 V/μm to reach a current density of 500 A/m^2.

Objective for study

Introduction: Example 12

Conventional imaging modalities, including X-ray and ultrasound imaging, have been reported with potential applications in biomedical fields. However, these are not sensitive tools for pre-clinical imaging in disease models, such as cancer detection and diagnosis, because they provide small contrast for imaging between normal and diseased tissue. This can be substantially improved by means of the fluorescence diffuse optical tomography (FDOT) method [1–5]. FDOT is an imaging modality that aims at reconstructing the 3-D distributions of fluorescent agents inside live small animals. In this technique, a fluorescent contrast agent is applied to provide further optical contrast in the target with respect to the background phantom. Because the blood vessels in cancers are leaky, the contrast agent tends to accumulate in cancers when the contrast agent is injected intravenously. The emitted fluorescence is then detected and reconstructed into 3-D fluorescence images in order to distinguish between normal and diseased tissue.

Highlight the recent trends

Although the FDOT method is expected to be a sensitive one, this has yet to be established. Therefore, such a trial should be carried out. Concerning the FDOT measurement, there are several optical parameters that affect the fluorescence temporal profile, for instance, absorption and scattering coefficients of materials, and target geometry. This paper intends to focus on the fluorescence temporal profile changes, relying on target position (centered to the near boundary). A time-correlated single photon counter served as a detection system. Firstly, fluorescence temporal profiles are presented as a function of geometrical configuration at the symmetrical point of indocyanine green target in tissue phantom. Next, an unknown target position was identified using the fluorescence peak intensity ratio. This analysis should be useful to decide the excitation-detection geometry for further aims in optical reconstruction imaging in a tissue model.

Need for study

Objective for study

Introduction: Example 13

Bimetallic nanoparticles can offer additional degrees of freedom when compared with the pure elemental particles, by altering physical properties of the latter. This can enable a wide range of applications, e.g. catalysis technologies [1–6] or optical devices [7–11]. Like bimetallic nanoparticles,

Highlight the recent trends

applications of bimetallic nanorods have served as a catalyst of one of the most active areas of the nanoscience. Bimetallic nanorods having large surface-to-volume ratios are among the nanomaterials used to improve selectivity and rate of metal-catalyzed reactions. Moreover, the bimetallic nanorods employed as nanocatalysts can provide a way for utilizing smaller amounts of expensive catalyst materials, by using less expensive metals as core materials. On the other hand, the structured core–shell nanocatalysts retain an efficient coupling surface plasmon resonance. They can, for instance, act as built-in sensing components, which are able to signal the exposure to biological agents and toxic chemicals before the doses of the latter become harmful.

Need for study

So far, the bimetallic nanorods have been used mainly in alloy or core–shell forms, relying significantly on the synthesis conditions, miscibility, and the kinetics of reduction of metal ions. Therefore, this study intends to observe the plasmonic properties of the core–shell particles where palladium-wrapped gold nanorods have been obtained with various palladium-shell thicknesses and surrounding media. Like platinum, palladium is unique for its catalytic properties. It is expected to be a highly useful industrial catalyst for producing hydrogen from methane, reducing automobile pollutant gases, and even in direct methanol-fuel cells.

Research objective

Gold has also attracted much attention as a potential core for bimetallic core–shell nanorods. The reasons are its strong optical absorption in the visible region [12,13] and good catalytic properties [14–16], which reveal clear size and shape dependences. Furthermore, the use of gold for the catalysis, instead of the other metals commonly employed for that aim, is more viable economically. In this respect our experimental results may prove to be very useful for the solar energy conversion by plasmonic luminescent solar concentrators. Moreover, the benefits from the plasmonic effect can be applied to develop metal-enhanced fluorescence via fluorescent dye-doped Au@Pd thin films. This will be the subject of our forthcoming article.

The impact of the present work concerns the two aspects. First, we have measured the exact values of the metal-component concentrations. Since the metal mass in the solution cannot be extracted from its precursors, one needs inductively coupled plasma-mass spectrometry examinations of the products obtained finally, especially in the case of our plasmonic investigations. We have used the mentioned spectrometric technique to measure the exact metal concentrations. Notice that some of the studies (see Refs. [17,18]) provide only approximate estimations issuing from the metal sources and the conversion efficiency. Although these rough concentration estimations are of the same order of magnitude, they can vary notably from the exact values. Second, we have built a simpler, lower-cost spectrometer, as compared with its commercial counterparts.

Introduction: Example 14

Recently, nanomaterials are important for many applications in different areas, such as electronic and optic nanodevices, biosensors, solar cells, and chemical catalysts, partly due to their size-dependent physical [1–3] and chemical properties [4–6]. For example, a broad range of the optically active semiconductor materials can serve as building blocks for miniaturized photonic and the optoelectronic nanodevices. Efforts will be made to assemble the nanostructures into useful architectures and practical nanodevices. Furthermore, one quite promising candidate is nanowire photovoltaics. Organic/inorganic photovoltaic nanodevices are the subject of intense research for low cost solar energy conversion. However such nanodevices are restricted in efficient charge transport. This is because of the highly discontinuous topologies of the donor–acceptor interface. Substituting the disordered inorganic phase with an aligned nanowire array is then suggested to improve the charge collection and the power conversion efficiency. Then, there are extensive researches on the development of new approaches to produce such nano-sized materials.

Nanomaterials including the metallic nanoparticles have been synthesized using the traditional methods, such as physical [7–12] and chemical methods [13–20]. For instance, self-assembly is an attractive alternative for nanolithography partly due to a high throughput and low cost. The self-assembled nanostructures are expected to show intrinsic bistability in the transport characteristics that will be exploited to realize a non-volatile quantum dot memory. They also show the signatures of Coulomb blockade and Coulomb staircase (at room temperature) in the quantum dots and the quantum wires, holding out the promise that these kinds of structures are suggested to find applications in the single-electron transistors including the other novel nanodevices. Nevertheless, there are some drawbacks, such as the

Show the recent trends

Need for research

technical simplicity, the low production yields, and contamination by a chemical reagent.

With a benefit of the pulsed-laser more recently, this study intends to report on a formation of zinc oxide nanoparticles using the laser ablation technique. We have suggested that pulsed laser ablation in the solution media will be an alternative for the production of the size-selected nanoparticles. These nanoparticles are expected to be a potential candidate for the photocatalyst and the semiconductor nanomaterials.

Research objective

On the other hand, since nanomaterials have gained increasing attention in various fields, including biomedical applications in cancer therapy, the scientific understanding of the interactions between nanomaterials with biological systems is important [21–29]. In recent times, zinc oxide nanoparticles are among the most commonly utilized nanomaterials in healthcare products, most notably in sunscreens, partly due to their superior efficiencies in ultraviolet absorption. Furthermore, zinc oxide nanoparticles offer a good biocompatibility with poor toxicity or even no toxicity in vitro and in vivo [20,21]. This reveals their excellent possibility of applications in life science. Therefore, zinc oxide nanoparticles may have a unique phototoxic effect upon the irradiation which can generate cytotoxic reactive oxygen species in aqueous media [22]. A cancer site irradiated with light possibly leads to a formation of the reactive oxygen species, resulting in the cell-killing.

Show the recent trends

Need for research

As the development of the cell-killing approach will give a powerful technique for some biological species for a distinction between various kinds of cancer cell lines, this study also intends to show that zinc oxide nanoparticles are good candidate nanomaterials for human breast cancer therapy. The MCF-7 cells served as the human breast cancer cells are treated by zinc oxide nanoparticles with different concentrations under ultraviolet irradiation in order to increase the cellular death. The typical characteristic phototoxic alters, for instance the chromatin condensation and fragmentation of the cell in the MCF-7 cells after the treatment with zinc oxide nanoparticles under ultraviolet irradiation, this is also discussed.

Research objective

Introduction: Example 15

Plasmonic-resonance properties of metallic nanoparticles have been extensively used to tailor the optical properties of organic molecules. Interactions of these nanoparticles with organic-dye molecules can become a basis for many applications, e.g. nanoscale lasers [1], surface-enhanced Raman spectroscopy [2–3], and plasmon-enhanced fluorescence spectroscopy [4]. These interactions can also play a key role in the photonic nanodevices. Metallic nanoparticles are able to either quench or enhance the intrinsic luminescence of an emitter, depending on the distance between the donor and the emitter. In general, when the incident light falls onto a surface of metallic nanoparticles, the enhancement of the fluorescence comes from the surface plasmon resonance. This plasmonic property results in increasing excitation decay rate and increasing radiative quantum efficiency. Moreover, the dipole energy accumulated at the surface of metallic nanoparticles can suppress the ratio of the radiative to non-radiative decay rates through the energy-transfer mechanism, thus leading to quenching of the intrinsic fluorescence from the dye molecules [5].

> Show the recent trends

Up to date, a number of studies have been reported on the properties of combinations of metallic nanoparticles with dyes. They have demonstrated that the metallic nanoparticles can affect the emission intensity of the dyes. Nonetheless, these works have mainly studied the effect of size and concentration of the nanoparticles. Although various shapes of metallic nanostructures have been dealt with in the past decade [6–8], the results concerned with introduction of the metallic nanoparticles into fluorescent materials are still limited. This implies a necessity for the further researches in this direction.

> Need for research

This study intends to focus on the effect of plasmonic interactions occurring in the metallic nanostructures of different shapes, namely gold–palladium core-shell nanorods (abbreviated further on as Au@Pd NRs), gold nanobipyramids (Au NBPs), and porous single-crystalline Pd nanocrystals (Pd PNCs), on the fluorescence of Rhodamine 6G (Rh6G) and Coumarin153 (C153) dyes. These dyes have been chosen because they are well enough studied and manifest good photostability

> Research objective

and high quantum yields [9,10]. In addition, they are widely used as fluorescent probes in various fields, e.g. in biology, chemistry, and medicine (see Refs. [11–13]). For photovoltaic applications, one can use the enhancement of fluorescence in these dyes, which takes place in thin solid films.

4.3 References

References are cited in the text with square brackets, such as [1]. If square brackets are not used, slashes may be used instead, such as /1/. Two or more references at a time may be put in a single set of brackets, such as [3,4]. The references should be numbered in the order in which they are cited in the text and should be listed at the end of the contribution under the heading "References." The reference style could be APA (American Psychological Association), Turabian, Harvard, MLA (Modern Language Association), Chicago, or Vancouver. However, most journals (i.e., Elsevier, Springer, John Wiley & Sons, Taylor & Francis, Oxford, Cambridge, etc.) have their own reference styles, and the reader can just follow their instructions. Also note that the reference styles will be different for Textbook, Book Chapter, Conference Proceeding, Patent, and the Internet.

4.3.1 Examples of References

References: Example 1 (according to Introduction 13)

[1] Liu H, Ye F, Yao Q, Cao H, Xie J, and Yang J. 2014. Stellated Ag-Pt bimetallic nanoparticles: An effective platform for catalytic activity tuning. *Sci. Rep.*, 4: 3669.

[2] Hu G, Nitze F, Gracia-Espino E, Ma J, Barzegar H R, Sharifi T, Jia X, Shchukarev A, Lu L, Ma C, and Wagberg T. 2014. Small palladium islands embedded in palladium–tungsten bimetallic nanoparticles form catalytic hotspots for oxygen reduction. *Nature Commun.*, 5: 5253.

[3] Liu W J, Qian T T, and Jiang H. 2014. Bimetallic Fe nanoparticles: Recent advances in synthesis and application in catalytic elimination of environmental pollutants. *Chem. Eng. J.*, 236: 448–463.

[4] Konig R Y G, Schwarze M, Schomacker R, and Stubenrauch C. 2014. Catalytic activity of mono- and bi-metallic nanoparticles synthesized via microemulsions. *Catalysts*, 4: 256–275.

[5] Rao V K and Radhakrishnan T P. 2013. Hollow bimetallic nanoparticles generated in situ inside a polymer thin film: Fabrication and catalytic application of silver–palladium–poly (vinyl alcohol). *J. Mater. Chem. A*, 1: 13612–13618.

[6] Zhang H, Haba M, Okumura M, Akita T, Hashimoto S, and Toshima N. 2013. Novel formation of Ag/Au bimetallic nanoparticles by physical mixture of monometallic nanoparticles in dispersions and their application to catalysts for aerobic glucose oxidation. *Langmuir*, 29: 10330–10339.

[7] Boote B W, Byun H, and Kim J H. 2014. Silver-gold bimetallic nanoparticles and their applications as optical materials. *J. Nanosci. Nanotechnol.*, 14: 1563–1577.

[8] Adekoya J A, Dare E O, Mesubi M A, and Revaprasadu N. 2014. Synthesis and characterization of optically active fractal seed mediated silver nickel bimetallic nanoparticles. *J. Mater.*, 2014: 184216.

[9] Perez J L J, Fuentes R G, Ramirez J F S, Vidal O U G, Tellez-Sanchez D E, Pacheco Z N C, Orea A C, and Garcia J A F. 2013. Nonlinear coefficient determination of Au/Pd bimetallic nanoparticles using Z-scan. *Adv. Nanoparticles*, 2: 223–228.

[10] Arquilliere P P, Santini C, Haumesser P H, and Aouine M. 2011. Synthesis of copper and copper-ruthenium nanoparticles in ionic liquids for the metallization of advanced interconnect structures. *ECS Transac.*, 35: 11–16.

[11] Sachan R, Yadavali S, Shirato N, Krishna H, Ramos V, Duscher G, Pennycook S J, Gangopadhyay A K, Garcia H, and Kalyanaraman R. 2012. Self-organized bimetallic Ag–Co nanoparticles with tunable localized surface plasmons showing high environmental stability and sensitivity. *Nanotechnology*, 23: 275604.

[12] Hu M, Chen J, Li Z Y, Au L, Hartland G V, Li X, Marquez M, and Xia Y. 2006. Gold nanostructures: Engineering their plasmonic properties for biomedical applications. *Chem. Soc. Rev.*, 35: 1084–1094.

[13] Dykman L A and Khlebstov N G. 2011. Gold nanoparticles in biology and medicine: Recent advances and prospects. *Acta Nature*, 3: 34–55.

[14] Gates B C. 2013. Supported gold catalysts: New properties offered by nanometer and sub-nanometer structures. *Chem. Commun.*, 49: 7876–7877.

[15] Zhou X, Xu W, Liu G, Panda D, and Chen P. 2010. Size-dependent catalytic activity and dynamics of gold nanoparticles at the single-molecule level. *J. Amer. Chem. Soc.*, 132: 138–146.

[16] Zhu Y, Jin R, and Sun Y. 2011. Atomically monodisperse gold nanoclusters catalysts with precise core-shell structure. *Catalysts*, 1: 3–17.

[17] Ruibin J. 2013. Plasmonic properties of bimetallic nanostructures and their applications in hydrogen sensing and chemical reactions. *Dissert. Abs. Inter.*, 75–06: 1–176.

[18] Hao J and Hui W. 2014. Controlled overgrowth of Pd on Au nanorods. *CrystEngComm*, 16: 9469–9477.

[19] Pradeep T. *A Textbook of Nanoscience and Nanotechnology*. New York: McGraw Hill (2012).

[20] Sandrock M L, Geiger F M, and Foss C A. 1999. Synthesis and second-harmonic generation studies of noncentrosymmetric gold nanostructures. *J. Phys. Chem. B*, 103: 2668–2673.

[21] Palik E D. *Handbook of Optical Constants of Solids*. New York: Academic Press (1991).

References: Example 2 (according to Introduction 14)

[1] G. Guerrero, J. G. Alauzun, M. Granier, D. Laurencin, and P. H. Mutin, *Dalton Trans.*, **42**, 12569 (2013).

[2] K. Janani, C. P. Kala, R. M. Hariharan, S. Sivasathya, and D. J. Thiruvadigal, *Austin J. Nanomed. Nanotechnol.*, **2**, 1 (2014).

[3] C. P. Li *et al.*, *Anal. Methods*, **6**, 1914 (2014).

[4] Y. Ogata and G. Mizutani, *Phys. Res. Int.*, **2012**, 969835/1 (2012).

[5] T. Iwai and G. Mizutani, *Phys. Rev. B*, **72**, 233406-1 (2005).

[6] K. Locharoenrat, A. Sugawara, S. Takase, H. Sano, and G. Mizutani, *Surf. Sci.*, **601**, 4449 (2007).

[7] N. A. Tuan and G. Mizutani, e-*J. Surf. Sci. Nanotechnol.*, **7**, 831 (2009).

[8] S. E. Lohse, J. A. Dahl, and J. E. Hutchison, *Langmuir*, **26**, 7504 (2010).

[9] S. Iravani, *Green Chem.*, **13**, 2638 (2011).

[10] C. Gutierrez Wing, J. J. Velazquez-Salazar, and M. Jose-Yacaman, *Methods Mol. Biol.*, **906**, 3 (2012).

[11] S. Y. Li and M. Wang, *Nano Life*, **2**, 1230002 (2012).

[12] S. Sinha, S. K. Chatterjee, J. Ghosh, and A. K. Meikap, *J. Appl. Phys.*, **113**, 1237041 (2013).

[13] H. Kawasaki, *Nanotechnol. Rev.*, **2**, 5 (2013).

[14] M. A. Mahmoud, D. O'Neil, and M. A. El-Sayed, *Chem. Mater.*, **26**, 44 (2014).

[15] J. Xu *et al.*, *Molecules*, **19**, 11465 (2014).

[16] B. Swathy, *IOSR J. Pharm.*, **4**, 38 (2014).

[17] C. N. R. Rao, S. R. C. Vivekchand, K. Biswas, and A. Govindaraj, *Dalton Trans.*, **11**, 3728 (2007).

[18] G. M. Herrera, A. C. Padilla, and S. P. Hernandez-Rivera, *Nanomaterials*, **3**, 158 (2013).

[19] A. M. Atta, H. A. Al-Lohedan, and A. O. Ezzat, *Molecules*, **19**, 6737 (2014).

[20] P. Saravanan, R. Gopalan, and V. Chandrasekaran, *Defence Sci. J.*, **58**, 504 (2008).

[21] M. E. Davis, Z. Chen, and D. M. Shin, *Nature Rev. Drug Discovery*, **7**, 771 (2008).

[22] S. Nazir, T. Hussain, U. Rashid, and A. J. MacRobert, *Nanomed. Nanotechnol. Biol. Med.*, **10**, 19 (2014).

[23] J. Sun, S. Wang, D. Zhao, L. Weng, and H. Liu, *Cell Biol. Toxicol.*, **27**, 333 (2011).

[24] W. Ye *et al.*, *Int. J. Nanomed.*, **7**, 2641 (2012).

[25] F. Tang and D. Chen, *Adv. Mater.*, **24**, 1504 (2012).

[26] G. Bao, S. Mitragotri, and S. Tong, *Annu. Rev. Biomed. Eng.*, **15**, 253 (2013).

[27] W. Lohcharoenkal, L. Wang, Y. C. Chen, and Y. Rojanasakul, *Biomed. Res. Int.*, **2014**, 180549 (2014).

[28] S. Mitragotri and P. Stayton, *MRS Bulletin*, **39**, 219 (2014).

[29] S. W. Morton *et al.*, *Sci. Signal*, **7**, 44 (2014).

[30] Joint Committee on Powder Diffraction Standards Diffraction Data File, JCPDS International Center for Diffraction Data No. 36-1451, 1991.

[31] C. Jagadish and S. Pearton, *Zinc Oxide Bulk, Thin Films and Nanostructures: Processing, Properties and Applications* (Elsevier, Amsterdam, 2007).

[32] X. W. Sun and Y. Yang, *ZnO Nanostructures and their Applications* (Pan Stanford, Singapore, 2012).

[33] N. Pante and M. Kann, *Mol. Biol. Cell*, **13**, 425 (2002).

[34] Y. Toduka, T. Toyooka, and Y. Ibuki, *Environ. Sci. Technol.*, **46**, 7629 (2012).

References: Example 3 (according to Introduction 15)

[1] Noginov M A, Zhu G, Belgrave A M, Bakker R, Shalaev V M, Narimanov E E, Stout S, Herz E, Suteewong T, and Wiesner U. 2009. Demonstration of a spaser-based nanolaser. *Nature*, 460: 1110–1112.

[2] MacLaughlin C M, Mullaithilaga N, Yang G, Ip S Y, Wang C, and Walker G C. 2013. Surface-enhanced Raman scattering dye-labeled Au nanoparticles for triplexed detection of Leukemia and Lymphoma cells and SERS flow cytometry. *Langmuir*, 29: 1908–1919.

[3] Lim D K, Jeon K S, Hwang J H, Kim H, Kwon S, Suh Y D, and Nam J M. 2011. Highly uniform and reproducible surface-enhanced Raman scattering from DNA-tailorable nanoparticles with 1-nm interior gap. *Nature Nanotechnol.*, 6: 452–460.

[4] Iosin M, Baldeck P, and Astilean S. 2009. Plasmon-enhanced fluorescence of dye molecules. *Nucl. Instrum. Meth. Phys. Res. B*, 267: 403–405.

[5] Kang K A, Wang J, Jasinski J B, and Achilefu S. 2011. Fluorescence manipulation by gold nanoparticles: From complete quenching to extensive enhancement. *J. Nanobiotechnol.*, 9: 16.

[6] Lee J H, Gibson K J, Chen G, and Weizmann Y. 2015. Bipyramid-templated synthesis of monodisperse anisotropic gold nanocrystals. *Nature Commun.*, 6: 7571.

[7] Henning A M, Watt J, Miedziak P J, Cheong S, Santonastaso M, Song M, Takeda Y, Kirkl A I, Taylor S H, and Tilley R D. 2013. Gold–palladium core–shell nanocrystals with size and shape control optimized for catalytic performance. *Ang. Chemie.*, 52: 1477–1480.

[8] Zhang J, Feng C, Deng Y, Liu L, Wu Y, Shen B, Zhong C, and Hu W. 2014. Shape-controlled synthesis of palladium single-crystalline nanoparticles: The effect of HCl oxidative etching and facet-dependent catalytic properties. *Chem. Mater.*, 26: 1213–1218.

[9] Kubin R F and Fletcher A N. 1982. Fluorescence quantum yields of some rhodamine dyes. *J. Lumin.*, 27: 455–462.

[10] Li H, Cai L, and Chen Z. *Advances in Chemical Sensors*. New York: InTech (2012).

[11] So H S, Rao B A, Hwang J, Yesudas K, and Son Y A. 2014. Synthesis of novel squaraine–bis(rhodamine-6G): A fluorescent chemosensor for the selective detection of Hg^{2+}. *Sensors and Actuators B: Chemical*, 202: 779–787.

[12] Zhang L, Wang J, Fan J, Guo K, and Peng X. 2011. A highly selective, fluorescent chemosensor for bioimaging of Fe^{3+}. *Bioorganic Medicinal Chem. Lett.*, 21: 5413–5416.

[13] Wagner B D. 2009. The use of Coumarins as environmentally-sensitive fluorescent probes of heterogeneous inclusion systems. *Molecules*, 14: 210–237.

4.4 Problems

Pb.1. Discuss the sources of literature review.

Pb.2. Explain the difference between theoretical and conceptual frameworks.

Pb.3. Select one topic of your interest and then show a statement of problems, write objective, and outcome of the study.

Pb.4. Read some references/sources and fill out the following table. Compare your findings with your colleagues.

Annotated literature for paper title "_____"

Author	Reference/Source	Findings
A		
B		
C		
D		
E		

Pb.5. Suppose you have referred one journal, write the reference style according to APA (American Psychological Association), Turabian, Harvard, MLA (Modern Language Association), Chicago, and Vancouver.

Pb.6. Give an example of the following references: textbook, book chapter, conference proceeding, patent, and the Internet.

Pb.7. Discuss the strong and weak points of the following introductions:

Introduction (A)

Copper is proposed as a valuable alternative for future interconnection applications because it carries electrical signals faster than aluminum [1]. One of its interesting characteristics is its exhibition of strong absorption peaks in visible spectrum due to surface plasmons when Cu wires' dimensions are reduced to nano-scale [2]. Although absorption is still the primary optical property of interest, other spectroscopic techniques including optical second harmonic generation (SHG) are also used to understand the anisotropic nature of these nanomaterials.

Kitahara et al. have studied SHG in Au nanowires on NaCl(110) substrate fabricated through a shadow deposition method [3]. They have reported that the SH response is weak when the electric field is perpendicular to the wire axes because of the cancellation of the incident field by the depolarization field. To extend our knowledge, we have measured the SH intensity from Cu nanowires on NaCl(110) template produced by the same fabrication technique [4]. We have found that the azimuthal angle and polarization dependence of the SH light in Cu nanowires is different from that of Au nanowires. The resonant enhancement of the SH intensity has been observed when the electric field component along the surface normal is enhanced by the plasmon-resonant coupling in the wires. The aim of this work is to address this issue by combining the results of the SH intensity patterns with the theoretical modeling based on finite-difference time-domain (FDTD) method.

Introduction (B)

Metal nanostructures have received a steadily growing interest as a result of their peculiar and fascinating properties as well as the applications superior to their bulk counterparts [1]. Generation of such nanometer-sized structures has shown a wealth of optical phenomena. For example, the dipolar mode leads to the strong absorption intensity at specific wavelengths for metal nanoparticles. It is responsible for the characteristic colors associated with suspensions of colloidal particles [2].

Among the nanometer structures, optical anisotropy has been studied for Au- and Ag-nanowires [3]. The excitation of dipolar plasmon mode dominates the extinction spectra for the polarization direction perpendicular to the wire axes. The spectral position of the plasmon resonance can be tuned by an appropriate choice of the nanowire geometry and material. This spectral selectivity may play a key component in the future optical devices. However, no detailed optical studies have been carried out for Cu nanowires.

In our previous reports, we chose Cu nanowires as a promising candidate for future interconnection applications and measured the azimuthal angle dependence of their second harmonic (SH) intensity in air at the fundamental photon energy of 1.17 eV. The nanowires of Cu were fabricated by a shadow deposition method on the NaCl (110) crystal facets [4]. The linear absorption spectra of Cu nanowires showed plasmon maxima for the incident electric field parallel to the long wire axes at a lower photon energy than for the perpendicular polarization [4].

We performed a phenomenological analysis of the SH intensity by fitting the theoretical patterns to the experiment assuming a Cs symmetric Cu nanowires/NaCl(110) system [5]. The contributions of the χ^2_{323}, χ^2_{311}, and χ^2_{222} elements dominated the SH signal for Cu nanowires. The numbers 1, 2, and 3 correspond to the [001], [1‾10], and [110] directions on the NaCl(110) surface, respectively. The χ^2_{222} element was judged to originate from the broken symmetry in the surface structures [5,6]. We have pointed out that there are two candidate origins of the χ^2_{311} and χ^2_{323} elements in Cu nanowires [5]. The first one is the variation of the electromagnetic field components along the surface normal coupled to the local plasmons in the nanowires. The second one is the lightning-rod effect. However, the understanding of these physical origins has not yet been gained well. The aim of this paper is to clarify this point by combining the results of the patterns of the SH intensity with a theoretical modeling based on finite-difference time-domain (FDTD) method.

Introduction (C)

Optical sum-frequency generation (SFG) is one of the lowest-order nonlinear optical processes in a noncentrosymmetric medium [1,2]. The microscopy using this nonlinear optical response has been used to monitor molecular vibrational images in biological materials [3]. The selective observation of asymmetric species in these materials cannot be achieved by other vibrational microscopies such as infrared and Raman microscopies [4,5]. The conventional SF microscopy has worked well only for thin specimens because they are more or less two dimensional and thus the whole material lies in the same focal plane [3]. However, when dealing with a thick specimen, this resolution is lost in conventional microscopy because only one thin slice in the specimen can be in focus. Most of the parts in the specimen are out-of-focus and the resulting image is rather out-of-focus.

In order to make a first possible step to overcome this drawback, we have constructed a confocal SF microscope and have performed a test measurement. Several vibrational microscopies have been developed to investigate molecular properties in complex systems. For instance, IR absorption and Raman microscopies have been applied to show distinction between molecular species using the spectroscopic information [4,5]. The observation of starch granules in a water plant has been performed by using a conventional SF microscope [3]. However, the development of a confocal SF microscope has not been reported. Therefore, to our knowledge, such a trial should be carried out.

In this experiment, a specimen of ZnS polycrystals was illuminated with two coherent light beams. One of the light beams (ω_{IR}) was tunable to the IR absorption bands of the sample. Another light beam provided visible light (ω_{VIS}). If the specimen has a noncentrosymmetric structure, the two beams can couple nonlinearly [1,2].

ZnS possess a crystal symmetry of a zincblende phase resulting in a noncentrosymmetric structure. The symmetry of this structure will therefore permit the second-order nonlinear phenomena including SFG from the bulk region of the crystal [1,2]. ZnS is of particular interest because it is an important device material for the detection, emission, and modulation of visible and near ultra violet light [6,7]. In particular, it is believed to be one of the most promising materials for blue light emitting laser diodes and thin film electroluminescent displays [8,9]. The advantages of the confocal SF microscopy compared with our previous conventional SF microscopy can be an improvement in spatial resolution and a very effective suppression of stray light from out-of-focus areas of the specimen under study [10]. Another important point is the function of optical sectioning of samples due to the small depth of the intense field formed in thick objects. In connection with these abilities, we can obtain images of a specimen that would otherwise appear blurred when viewed with a conventional microscopy.

Introduction (D)

The microscopy using optical sum-frequency generation (SFG) [1,2] has been utilized to monitor molecular vibrational images of materials with non-centrosymmetric structures [3,4]. So far several vibrational microscopies have been developed to investigate molecular properties in complex systems. For instance, IR absorption and Raman microscopies have been applied to show distinction between molecular species using the spectroscopic information [5,6]. The observation of starch granules in a water plant has been performed by using a conventional SF microscope [4]. However, the development of a SF microscope equipped with confocal optics has not been reported yet except for our group [7]. Therefore, to our knowledge, such a trial should be carried out.

In this study we have constructed an optical sum frequency microscope with confocal optics [7]. In the demonstration of this SFM, a specimen of ZnS polycrystals was illuminated with two coherent light beam pulses. One of the light beams (ω_R) was tunable to the IR absorption bands of the sample. In this study it is tentatively set to 2890 cm^{-1} in the CH stretching region for future application to biological systems. Another light beam provided visible light (frequency _VIS). If the specimen has a non-centrosymmetric structure, the two beams can couple nonlinearly [1,2]. ZnS possesses a crystal symmetry of a zincblende phase resulting in a non-centrosymmetric structure. The symmetry of this structure will therefore permit the second-order nonlinear phenomena including SFG from the bulk region of the crystal [1,2].

The advantages of the confocal SF microscopy compared with previous conventional SF microscopes can be an improvement in its spatial resolution [4]. Another important point is the function of optical sectioning of samples due to the small depth of the intense field formed in thick objects.

Introduction (E)

Applications in nanodevices demand the large areas with high throughput of the metallic nanowires. A shadow deposition method is considered as one of the promising procedures to fulfill this requirement because this self-organized pattern formation is completed without breaking vacuum [1–3].

With the development of pulsed laser technology, nonlinear optical techniques including second harmonic generation (SHG) are used to understand the anisotropic nature of those nanomaterials [4–6]. Owing to its intrinsic surface sensitivity, the resonant coupling of surface plasmons in the metallic nanowires with SHG might result in an enhancement of the electromagnetic field. The linear dielectric response of the composites of the metallic nanowires in the dielectric medium might lead to an enhancement of second order optical nonlinearity. These parameters motivate us for further studies to gain inside into a microscopic origin of nonlinearity of metallic nanostructures.

Introduction (F)

There are many potential applications for which nanowires may become important: for electronic and optoelectronic nanodevices, for metallic interconnects of quantum nanodevices. These applications in any devices demand the large areas with high throughput of the metallic nanowires. A shadow deposition method is one of the promising procedures to fulfill this requirement. In contrast to a lithographic method [1], this self-organized pattern formation could be also completed without breaking vacuum [2].

With the development of pulsed laser technology, nonlinear optical techniques including second harmonic generation (SHG) are used to understand the anisotropic nature of nanowires [3]. A great advantage of these studies over linear optical techniques is its intrinsic surface sensitivity due to the fact that SH signal originates from a thin interfacial region [4–6]. Owing to this surface sensitivity, the resonant coupling of surface plasmons in nanowires with SHG might result in an enhancement of the electromagnetic field. These factors motivate the need for further studies toward the elucidation of the microscopic origin of the nonlinearity of metallic nanostructures.

Introduction (G)

Interests in nanotechnology have inspired a lot of researches on metal nanowires and opened up a new prospect of applications in future electronic and optical devices [1,2]. Cu is of particular interest because it can be suitable for potential application such as in interconnection in electronic circuits. The developments of fabrication approaches to produce Cu nanowires have been reported in literatures [3–5]. For optical application requiring nanowire arrays of large areas with a high throughput, shadow deposition technique by templating against relief structures on the (110) surfaces of NaCl crystals has been proposed [6].

So far, many optical studies on metal nanowires have been performed. Sandrock et al. have disclosed local-field enhancement of noncentrosymmetric Au nanostructures [7], Schider et al. have shown excitation of dipolar plasmon mode of Au and Ag [2], and Kitahara et al. have observed optical nonlinearlity from Au nanowire arrays [8].

Despite a number of researches on bulk properties of Cu films [9–11], no literature can be found on optical properties of Cu nanowire arrays and hence there is a need here to focus on such nanowire materials.

Introduction (H)

Fluorescence Diffuse Optical Tomography (FDOT) is an emerging imaging modality with potential application in biomedical field.[1–3] The technique is considered to obtain quantitative images of fluorescence, as well as, absorption and scattering in diffuse media, such as tissue.[4,5] This approach relies on the presence of fluorophores in tissue that emit fluorescence light after illumination by near infrared light source. Although this method is expected to be very sensitive one, the technique has not been established yet.

The resolution of the optical reconstruction images obtained from FDOT method depends on several parameters, such as absorption and scattering coefficients of materials, and target geometry. In this contribution, we focus on the fluorescence temporal profile change depending on the geometry, such as, the target size. We use a phantom as a physical tissue model for further study of tomographic images of tissue samples in our future works. We are going to present the latest results on the fluorescence temporal profiles as a function of geometrical configurations.

Introduction (I)

In the recent years, scientific interest in the metal nanostructured surfaces in various forms has given birth to the new research field of nanoplasmonics, which is increasing very rapidly and is believed to lead to new materials whose optical properties can be engineered for various applications, for instance, for phototherapy based on the nanorods [1], nanoplasmonic counterpart to laser is proposed as a nanoscale quantum generator of nanolocalized coherent intense optical fields [2], for plasmon-assisted solar energy conversion [3], and for nanoplasmonic circular polarizers or optical superlenses [4].

So far, the primary role of the metallic nanostructured surfaces has shown a wealth of optical phenomena. For example, the external light photons are coupled with the electrostatic field of plasmons creating the strong absorption band [5]. With the development of the laser technology, nonlinear optical method including second harmonic generation (SHG) is also used to gain insight into the anisotropic nature of these nanomaterials. A great merit of these studies over the linear optical method is its intrinsic surface sensitivity partly due to the SH signal originated from the thin interfacial region [6–8]. Owing to this surface sensitivity, the resonant coupling of surface plasmons in the metallic nanostructured surfaces with SHG might result in the field enhancement. The linear dielectric response of the composites of the metallic nanostructured surfaces in the dielectric medium might lead to an enhancement of second order optical nonlinearity. These factors motivate us for further investigation in order to understand the microscopic origin of the nonlinearity of metallic nanostructured surfaces.

In this work, we observe and compare the second-order nonlinear optical properties of arrays of metal nanowires supported by the alkali halides template.

Chapter 5

Data Collection Methods

Data collection method is the second major part of a paper. In this chapter, you will see how to present a detailed methodology, which readers can replicate to check your study. The methodology can be presented in terms of experiment or design and construction or even both.

5.1 Experiment Method

The methodology identifies the type of experimental study. You should normally use past tense and active voice in this section. First, you show what the sample is, how to ensure ethics in research with animal or human subjects (e.g., permission, etc.), and what instruments to be used. You should state the name of the instruments, validity, and reliability. Then you show the data collection procedure for the study. Finally, you propose analysis of data based on your research questions or hypothesis.

5.1.1 Examples of Experiment Method

Experiment Method: Example 1

NaCl(110) single crystals (miscut angle <1°) having dimensions 10 × 18 × 4 mm^3 were dipped in distilled water for 10 sec and in ethanol

Research Methodologies for Beginners
Kitsakorn Locharoenrat
Copyright © 2017 Pan Stanford Publishing Pte. Ltd.
ISBN 978-981-4745-39-0 (Hardcover), 978-1-315-36456-8 (eBook)
www.panstanford.com

for 1 min in two cycles before being mounted on the substrate heating stage in an ultrahigh vacuum deposition chamber. The 200–300 nm thick homoepitaxial layers of NaCl were deposited after the substrate annealing at 200–400°C. The 10 nm thick SiO layers were then deposited as passivation layers. Cu was deposited at the rate of 0.6 nm/min onto the substrate at the flux angle of 65° with respect to the surface normal to form nanowires or nanodots. On the arrays of Cu nanowires and nanodots, 10 nm thick SiO layers were deposited as protection layers. For transmission electron microscopy (TEM) observation, Cu nanowire arrays sandwiched by SiO layers were separated from the NaCl substrates by dissolving the substrates in distilled water, and mounted on 200 mesh Cu grids.

Absorption spectra at the normal incidence were measured as a function of the sample rotation angle. These spectra were recorded in the wavelength 250–1100 nm by a spectrometer equipped with a polarizer accessory, a Xe lamp source, and a photomultiplier tube.

Experiment Method: Example 2

Cu nanowires were fabricated through shadow deposition technique in UHV of approximately 2×10^{-9} Torr [8]. Faceted NaCl(110) substrates were prepared as a nanoscale template by water etching and annealing at 200–400°C for 2–4 h. The regular arrays of Cu nanowires were then deposited from a crucible aligned by 65° from the template normal on the faceted surfaces at room temperature, and they were sandwiched by 10 nm thick SiO layers. By removing the NaCl template, we checked the morphology of the nanowire arrays by transmission electron microscopy (Hitachi: HF-2000) at an accelerating voltage of 200 kV. Fabrication parameters and sample descriptions are shown in Table 1.

Table 1   Deposition parameters and geometrical of the fabricated nanowires

Sample code	Nominal thickness (nm)	Wire width (nm)	Periodicity (nm)	Homoepitaxial growth temperature (°C)
C7	20	17	27	200
C8	10	14	27	200
C9	20	19	40	400
C10	10	15	40	400

To probe the plasmon resonances, absorption spectra at the normal incidence as a function of the sample rotation angle are recorded for the wavelength 250–1100 nm by a spectrometer equipped with a polarizer accessory, a Xe lamp source, and a photomultiplier tube.

Experiment Method: Example 3

Samples with varying wire widths of Cu nanowires on the faceted NaCl (110) substrates were fabricated by a shadow deposition method [2]. Cu nanodots were produced by the same process.

The probe-light pulses with a fundamental photon energy of 1.17 eV was generated by a mode-locked Nd:YAG laser. The pulse energy was set at 50 mJ/pulse. The fundamental light beam was focused onto the sample surface with a spot diameter of 2–3 mm and the incident angle was 45°. The SH signal was detected by a photomultiplier and the absolute SH intensities were obtained by normalizing the SH response to that of the α-SiO$_2$ (0001). To measure the azimuthal angle dependence of the SH intensity, the samples were mounted on an automatic rotation stage with the surface normals set parallel to the rotation axis of the stage. The measurements were carried out in four different input/output polarization combinations: p-in/p-out, p-in/s-out, s-in/p-out, and s-in/s-out. All SHG observations were performed in air at room temperature.

Experiment Method: Example 4

The arrays of Cu nanodots and those of Cu nanowires were prepared by a shadow deposition of Cu atoms on a faceted NaCl(110) template. The sample preparation was described in our previous paper [5].

For SHG measurements, the light source of the fundamental photon energy from 1.2 to 2.3 eV was an optical parametric generator and amplifier system driven by a mode-locked Nd:YAG laser. The excitation beam was then focused on the sample surface at the incident angle of 45°. The SH signal from the sample was detected by a photomultiplier. Finally, the absolute magnitudes were obtained by normalizing the SH response to that of the α-SiO$_2$ (0001).

Experiment Method: Example 5

The measurement system consists in a picosecond laser system and a time-resolved detection system. Briefly, the excitation laser was generated by a picosecond Ti:Sapphire laser system (Spectra Physics) and a pulse picker system (Cono Optics). The laser was coupled to a graded index multi-mode fiber and injected into the cylinder sample from the position at 0, 90, and 180° from the target plane. The emission or scattering light was collected by 3 mm bundled fibers at the other positions at 30° each and delivered to a fiber switch. The fiber switch selected one fiber of them and finally the signal reached the detection system. A time-correlated single photon counting (TCSPC) system (Hamamatsu TRSF-20) was employed for the detection system. A computer controlled the system and accumulated the counting data and temporal data. The fluorescence was measured with a 5-sheets film filter (Fuji Film, IR820). We used 780 nm for excitation and the ND filters were used to maintain the appropriate counting rate for TCSPC system in general measurements.

The tissue mimic phantoms were made by a 3 cm polyoxyemethylene (POM) resin cylinder. The phantom has a small target hole and the fluorophore solution filled the hole. In our standard measurements, the target size was 6 mm (diameter) at the middle of the radius of the 3 cm cylinder (at 7.5 mm from surface) and the fluorescence target consisted in 1 µM indocyanine green (ICG) in 1% Intralipid®.

Experiment Method: Example 6

The block diagram of the experimental setup is shown in Fig. 1(A). A Ti:Sapphire laser (Tsunami, Spectra Physics) with a pulse picker system (Cono Optics) generated a picosecond pulse train at 780 nm for the time–domain measurement. The laser was coupled to a graded index multi-mode fiber and injected into the sample. The emission and excitation light were collected by 3 mm bundled fibers at appropriate positions around the sample. Each fiber was selected by a fiber switcher and then the detected signal was delivered to a multi-channel plate photomultiplier tube (MCP-PMT, R3809U-61, Hamamatsu). A time-correlated single photon counting (TCSPC) system (Hamamatsu TRSF-20S) yields the temporal profile of the excitation light or emission light. The fluorescence was selected by an interference filter (D850/40m, Chroma). Neutral density filters were used to maintain the appropriate counting rate for TCSPC system in the measurements.

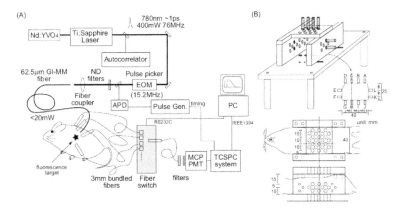

Figure 1 Time–domain measurement. (A) The schematic diagram of the instrument. (B) The schematic of the animal holder and the animal position.

All animal experiments were approved by the institutional animal care and use committee of Hokkaido University (#09-0016). Rats (male Wistar rat about 8 weeks old) were anesthetized by intraperitoneal injection of 1.1 g/kg b.w. Urethan. First, the hair was shorten by scissors and then removed by using depilatory cream to expose the skin for the measurement. Then, about few mm incision in the abdomen at a lower midline was made and the abdominal cavity was opened to implant a glass capillary (1.6 mm I.D. and 3–4 cm length), which contains a 2 µM indocyanine green (ICG) and 1% Intralipid mixture solution. The incision was closed by a surgical string after the implantation of the capillary. The one side of the capillary was glued to the surgical string and fixed to the skin to minimize the motion inside. At the end of the experiments, the rat was euthanized by an overdose of pentobarbital.

The schematic illustration of the animal holder is shown in Fig. 1(B). One side is the moving wall to make appropriate contact of the optical fibers to the rat. The animal was placed on its back in the holder. The optical fibers were placed to touch the body around the middle of the target capillary to measure the cross section of the rat. Decay curves of the excitation and emission light were measured with 50 pairs of the excitation and detection points for the reconstruction. The decay curves were accumulated up to 60 sec except some of the measurement points without the target.

Experiment Method: Example 7

The AlN films on the quartz substrates were deposited using aluminum target (99.9%) of 3 inches in diameter. The depositions performed in our home-made RF magnetron sputtering system. The base pressure in our vacuum chamber system was at approximately 2.0×10^{-6} mbar. The shutter was firstly needed to close in order to protect the substrates. For pre-sputtering, argon gas was then introduced into a chamber for 30 minutes in order to clean the target. The pressure at this time was about 2.3×10^{-3} mbar. The power of RF was set at 200 W. Next, nitrogen concentrations (CN) were varied at 20%, 40%, 60%, 80%, and 100% to produce five different film samples. For example, 20% CN was used to prepare one film sample. All film thicknesses were controlled at about 1 μm. Finally, the films were cooled down under argon atmosphere for 60 min to reach room temperature.

The crystal structure of the films was studied by X-ray diffraction (XRD; Bruker D8 Advanced). The surface morphology was investigated by atomic force microscopy (AFM). To determine the optical properties and thicknesses of the deposited films on the quartz substrates, transmittance spectra at normal incidence were measured at wavelength range from 200 nm to 900 nm by a spectrophotometer (T90+ UV/VIS Spectrometer PG Instruments Ltd.). The film thicknesses determined by the transmittance spectra were confirmed by a field emission scanning electron microscope (FE-SEM). In addition, the Al sputtered atom that effect on the film properties was monitored by optical emission spectroscopy (OES: Ocean Optics USB4000).

Experiment Method: Example 8

The AlN films on the quartz substrates were fabricated by our home-made radio frequency reactive magnetron sputtering system under the nitrogen/argon mixture atmosphere. First, the substrates were ultrasonically cleaned by acetone, methanol, and isopropanol. After that we baked the substrates to remove the moisture. Next, the sputtering system was evacuated to achieve a base pressure of 2.0×10^{-6} mbar. The shutter was then closed to protect the substrates. The aluminum (99.9% purity) target was pre-sputtered by pure argon for 20 minutes. The pressure at this stage was about 2.3×10^{-3} mbar. The power of radio frequency was set at 200 W. After that, the nitrogen concentrations (CN) were altered at 20%, 40%, 60%, 80%, and 100% to produce five different film samples. Finally, the films were cooled down under argon atmosphere for 60 minutes to reach a room temperature.

The obtained films were analyzed as follows. Formation of Al-N bonding was confirmed by FTIR (Thermo Nicolet6700) and Raman spectroscopy (Renishaw InVia Raman Microscope). Crystal orientation was identified by 2θ scan of X-ray diffraction (XRD; Bruker D8 Advanced). Surface morphology was investigated by atomic force microscopy (AFM; Seiko SPA400). Columnar structure was observed by field emission scanning electron microscope (FE-SEM; Hitachi S-4700). To evaluate the electrical resistivity and piezoelectric properties, the metal–insulator–metal (MIM) structures were formed. First, a square-aluminum electrode was deposited on a quartz substrate by the thermal evaporation technique, and then AlN film was deposited on the square aluminum electrode by the same sputtering process. After that, another square-aluminum electrode was evaporated on the top of the film. The direct current resistivity was calculated from the current–voltage measurement (Keithley; 6485 picoampmeter and 230 programmable voltage source), while the piezoelectric coefficient was measured by a homodyne Michelson interferometer technique.

Experiment Method: Example 9

Each tip was prepared by immersing several millimeters of the lower end of a tungsten wire of 0.25 mm in diameter into an aqueous 1 M NaOH etching solution and applying a measured voltage between the wire and a stainless steel counter-electrode. The wire was cut into pieces about 30 mm in length, and then ultrasonically cleaned by methanol prior to etching. The etching was continued until the submerged portion of the wire dropped off into a beaker, leaving the usable tip suspended near the air–solution interface. During the electrochemical process, we used a home-made etching process controller to terminate the etching current on a microsecond timescale in order to optimize the tip shape. In principle, a drop-off method produced two tips simultaneously: the upper tip in the air and the lower tip in the solution. The total tip length in the electrolyte could not be long enough because the weight of the lower tip affected the drop-off process. Together with the difficulties to collect the dropped tip in the electrolyte, these disadvantages made us give up using the lower tip and we used the upper tip instead. The morphology of a produced single tip was finally observed by means of scanning electron microscopy (SEM). Its reproducibility was also confirmed by SEM.

For observation of field emission pattern, the fabricated tip was introduced into an ultrahigh vacuum (UHV) chamber with a base pressure of approximately 10^{-8} Torr at room temperature as shown in Fig. 1(a). A flat stainless-steel anode was placed at about 100 μm from the tip cathode. Light pulses from a Ti:Sapphire chirped pulse amplifier system with a center wavelength of 800 nm and a duration of 20 fs entered through a thin ZnSe window and were focused onto the tip as shown in Fig. 1(b). The tip was oriented perpendicular to the laser beam direction. The emission current as a function of the applied voltage, both with and without the laser illumination, was then monitored with a Keithley 6485 picoammeter. The experiments were performed for the tip bias voltages between 0 and 1000 V.

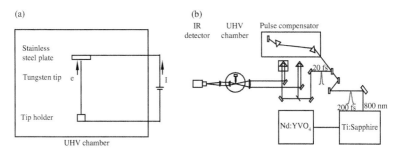

Figure 1 Field-emission setup of the tungsten nanotip; (a) inside the ultrahigh vacuum (UHV) chamber and (b) in a whole system. The distance between apex of the tip and stainless steel (SS) plate was about 100 μm.

Experiment Method: Example 10

A fluorescence target of 6 mm in diameter was performed with different target positions (4.5, 7.5, or 15 mm, apart from phantom surface) and concentrations (0.5, 1.0, and 2.0 mM indocyanine green dye in 1% Intralipid solution). This target was filled in a tissue phantom in which the top view is shown in Fig. 1. The target was then excited by incident light at position A (90°). The light source was a Ti:Sapphire laser with a central wavelength of 780 nm. Fluorescence light was detected by a photomultiplier. Detection points were focused at the symmetrical point B (30°) and D (−30°), as shown in Fig. 1.

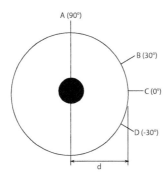

Figure 1 Coordination system of indocyanine green (ICG) target in tissue phantom of 60 mm in length and 30 mm in diameter. Target position (d) was 4.5, 7.5, or 15 mm, measured from the phantom surface to the target center. The black and white circles represent the ICG target and tissue phantom, respectively. The target size was 6 mm in diameter. The excitation point was defined at position A (90°). The detection point B and D was 30° and −30°, respectively.

Experiment Method: Example 11

A gold core was prepared from the solution of cetyltrimethylammonium bromide, using a seed mediated growth technique [19]. Palladium chloride components of different concentrations were then added to the gold core in order to form a palladium shell of varying thickness. The exact concentrations were examined with the inductively coupled plasma-mass spectrometry, which is usually used for determining the exact amounts of metals. The gold-core sizes and the palladium shell thicknesses were characterized using a transmission electron microscopy. The samples were centrifuged and redispersed in deionized water before the optical absorption measurements.

The absorption spectra (in arbitrary units) of the gold–palladium core–shell nanorods with different weight ratios of palladium and gold (Pd:Au = 0.20:1, 0.10:1, 0.05:1, and 0:1) were measured in the wavelength region 400–900 nm, using a home-made spectrometer equipped with polarizing accessories, a xenon lamp Hamamatsu L2273, and a photomultiplier Hamamatsu R585 (see Fig. 1). We used two monochromators in a series, which functioned as a double filter for rejecting any stray light. The exit slit of the first monochromator was the entrance slit for the second one. The absolute values of sample absorptions were obtained by normalizing the sample absorption response corresponding to plasmon resonances to that of a quartz reference plate.

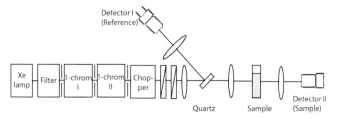

Figure 1 Schematic diagram of the optical setup.

Finally, the optical absorptions of a set of palladium-coated gold nanorods placed in the liquids with different dielectric constants (see Table 1) were studied.

Table 1 Chemical media with variable refractive indices

Ethanol fraction (%)	Toluene fraction (%)	Refractive index
0	0	1.33 (DI water)
100	0	1.36
67	33	1.41
50	50	1.43
33	67	1.45
0	100	1.49

The plasmon modes were also detected for the reference cases of refined ethanol and toluene. The experimental data was presented in terms of dependence of the plasmon-peak wavelength on the refractive index.

Experiment Method: Example 12

A zinc metal plate (purity of 99.99%) was placed in the quartz cell filled with 5 mL, 1 mM of an aqueous solution of the sodium dodecyl sulfate as a surfactant at the room temperature (Fig. 1). The laser light with the fundamental photon energy of 1.17 eV was originated by a Nd:YAG laser with the pulse energy set at 100 mJ per pulse. The beam was focused onto the surface of the target with the spot size of 2–3 mm in diameter. The liquid thickness was 1 mm beyond the target surface. The solution was stirred slowly during the ablation. The ablation times were varied from 10 to 20 min. The big particles and the free sodium dodecyl sulfate particles existing as the residues were finally removed from the solution using a centrifugal pump.

The optical properties and morphology of the prepared zinc oxide nanoparticles were investigated as follows. For the transmission electron microscope (TEM) observation, a drop of the solution of the zinc oxide nanoparticles was kept onto the copper grid and left in a vacuum oven at 50°C for 6 h. The produced specimen was then cleaned by distilled water in order to remove the free sodium dodecyl sulfate particles and put onto a sample stage of TEM. On the other hand, the crystal structure of the zinc phase was examined by X-ray diffraction (XRD). Finally, the absorption spectrum of the prepared solutions was recorded by UV–Vis spectrophotometer in the wavelength of 400–1100 nm.

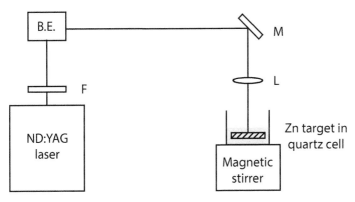

Figure 1 Block diagram of the experimental setup for producing the zinc oxide nanoparticles. The B.E., F, M, and L stand for beam elevator, filter, mirror, and lens, respectively.

The average crystalline size of the zinc oxide nanoparticles was approximately 45–60 nm. The human breast cancer cells (MCF-7, obtained from National Cancer Institute of Thailand) were cultured in completed cell culture medium (DMEM supplied with 10% fetal bovine serum and 1% penicillin-streptomycin) under the standard culture conditions (37°C with 5% carbon dioxide).

For the application of zinc oxide nanoparticles, the MCF-7 cells in mid-log phase at approximately 1.0×10^5 cells/mL, or 1 mL in the 24-well plate formats were left to attach onto the bottom surface of the well plates for 24 h before the nanoparticle exposure. Due to the superior efficiency of zinc oxide in absorbing UV (band gap of 3.37 eV), upon UV illumination, zinc oxide photocatalyst in aqueous media can produce the reactive oxygen species (ROS) for instance superoxide and hydrogen peroxide. The MCF-7 cells were then treated with a serial concentration of the zinc oxide nanoparticles (0 = the controlled cells, 2.5, 5, 10, 20, 40, 80, 160, 320, 640, 1280, 2560 µg/mL) under the ultraviolet irradiation in the ultraviolet chamber (UVC at wavelength of 254 nm) for another 24 h. Iron (II) chloride tetrahydrate and 3-(2-pyridyl)-5,6-diphenyl-1,2,4-triazine-4′,4″-disulfonic acid sodium salt were filled after the treatment. The zinc oxide nanoparticles at different concentrations were prepared after they were centrifuged and redispersed in different amount of DI water. The UV–Vis absorption spectroscopy was then performed in the wavelength of 400–700 nm under the room temperature of about 20°C. The ferrous ion chelating (%) was then calculated as follows: ferrous ion chelating (%) = [1 – (absorbance of the tested cells/absorbance of the controlled cells)] × 100. Each sample was carried out in five independent tests and the mean values were reported. It was noted that the zinc oxide nanoparticles offer a good biocompatibility with poor toxicity or even no toxicity in vitro and in vivo [20,21], we skip to test for the cell viability assay (MTT).

In order to confirm the presence of cell killing effect, we will check the morphological changes of the cells as follows. The MCF-7 cells upon the treatment were washed by phosphate buffer saline (PBS with 0.1 M at pH = 7.4). The MCF-7 cells were detached by trypsinization technique using 0.25% typsin/EDTA for 5–10 min. We then discarded the supernatant. The cells were resuspended with fresh media. The 400 µL cell suspension was transferred to microcentrifuge tube and filled with 100 µL of 0.4% trypan blue. The specimen was loaded onto the hemacytometer to quantify cell amounts. The staining images were also qualitatively observed using the phase contrast light microscope. The cell viability (%) was thus expressed as the following: cell viability (%) = (amount of the tested cells/amount of the controlled cells) × 100. Each sample was tested in at least five independent experiments and the mean values were reported.

Experiment Method: Example 13

The Rhodamine 6G and Coumarin153 were obtained from Aldrich (USA). Gold–palladium core-shell nanorods (Au@Pd NR), gold nanobipyramids (Au NBP), and single-crystalline porous Pd nanocrystals (Pd PNC) were purchased from Nanoseedz Limited (Hong Kong). These metallic nanostructures were chosen to investigate the influence of them on the fluorescence emission of the chosen dyes. This was because they could show the absorbance spectra overlapping with the fluorescent spectra of the chosen dyes as shown in Fig. 1. In our present study, three sets of the solutions (A, B, and C) were well prepared and they were exhibited in Tables 1–3.

Table 1 Set A (M = Molar)

Sample	Reference (Rh6G in DI) M	Sample (Rh6G + Au@Pd NR)	
		Rh6G (M)	Au@Pd (M)
A1	1.50×10^{-6}	1.50×10^{-6}	3.05×10^{-3}
A2	7.50×10^{-7}	7.50×10^{-7}	15.25×10^{-4}
A3	3.75×10^{-7}	3.75×10^{-7}	76.25×10^{-5}
A4	1.88×10^{-7}	1.88×10^{-7}	38.12×10^{-5}
A5	9.38×10^{-8}	9.38×10^{-8}	19.06×10^{-5}

Table 2 Set B

Sample	Reference (Rh6G in DI) M	Sample (Rh6G + Au NBP)	
		Rh6G (M)	Au (M)
B1	1.00×10^{-6}	1.00×10^{-6}	83.8×10^{-3}
B2	0.50×10^{-6}	0.50×10^{-6}	41.9×10^{-3}
B3	2.50×10^{-7}	2.50×10^{-7}	20.9×10^{-3}
B4	1.25×10^{-7}	1.25×10^{-7}	10.5×10^{-3}
B5	6.25×10^{-8}	6.25×10^{-8}	5.24×10^{-3}

Table 3 Set C

Sample	Reference (Cou153 in Ethanol) M	Sample (Cou153 + Pd PNC) Cou153 (M)	Pd (M)
C1	270.83×10^{-8}	270.83×10^{-8}	14.68×10^{-3}
C2	268.75×10^{-8}	268.75×10^{-8}	7.34×10^{-3}
C3	267.70×10^{-8}	267.70×10^{-8}	3.67×10^{-3}
C4	267.19×10^{-8}	267.19×10^{-8}	1.83×10^{-3}
C5	266.92×10^{-8}	266.92×10^{-8}	0.91×10^{-3}

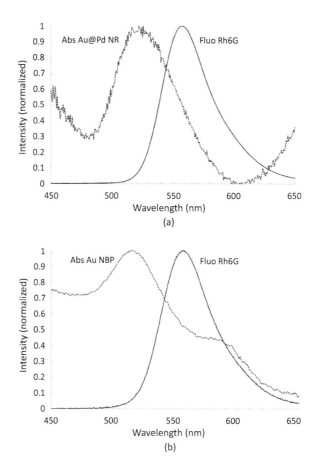

(a)

(b)

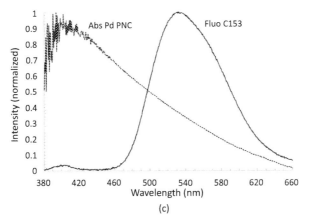

(c)

Figure 1 Absorption spectra of metallic nanostructures (Au@Pd NR, Au NBP, and Pd PNC) and fluorescence spectra of dyes (Rh6G or C153): (a) Absorption spectrum of Au@Pd nanorods and fluorescence spectrum of Rh6G; (b) Absorption spectrum of Au nanobipyramids and fluorescence spectrum of Rh6G; (c) Absorption spectrum of single-crystalline porous Pd nanocrystals and fluorescence spectrum of C153.

The absorption spectra of the prepared samples were measured at the wavelength of 350–800 nm by using the optical spectrometer (Avantes 2048). The tungsten lamp was used as a light source. In order to measure the fluorescence spectra of the dyes (Rh6G or C153), the high power green light emitting diodes (green LEDs at the excitation wavelength of 516 nm, 1 W power) were used as the excitation light for Rh6G, whilst ultraviolet LEDs at the excitation wavelength of 398 nm with 1 W power were served as the excitation light of C153. The right angle detection was employed as shown in Fig. 2.

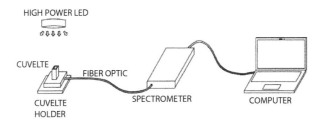

Figure 2 Schematic diagram of fluorescence spectra measurement by using the right angle detection.

5.2 Design and Construction

In addition to experiment method, the methodology can be of the type of general design and/or construction. First you show what the specimens are and what materials will be designed and constructed. Then you show the data collection procedures for the study.

5.2.1 Examples of Design and Construction

Design and Construction: Example 1

The optical delay line consists of a retro-reflector, an inclined reflection mirror, and a scanning mirror, as shown in Fig. 1(a).

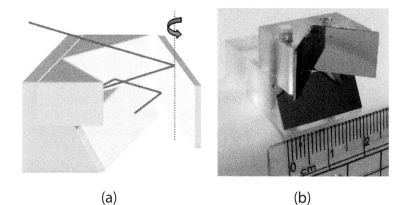

(a)   (b)

Figure 1 (a) Design of the optical delay line. (b) Photograph of the retro-reflector assembled with the reflection mirror.

The retro-reflector is fabricated with two right-angled prisms assembled on an aluminum jig where a reflection mirror inclined at 60 degrees is integrated. Figure 1(b) shows the photograph of the retro-reflector assembled with the reflection mirror where one can see that the size of the entire system is within 2 cm × 2 cm. As shown in Fig. 1(a), the light beam is incident at 30 degrees downward on the scanning mirror. After reflecting by the scanning mirror and the retro-reflector, the beam is reflected by the scanning mirror again. Finally, the beam is normally incident on the inclined reflection mirror such that the reflected beam is collinear with the incident beam. Since the reflected beam is guaranteed to be collinear with the incident beam regardless the tilted angle of the scanning mirror, no walk-off problem will be generated in such an optical delay line.

Figure 2 shows the propagating path of a light beam in the optical delay line where the tilted angle of the scanning mirror is assumed to be θ.

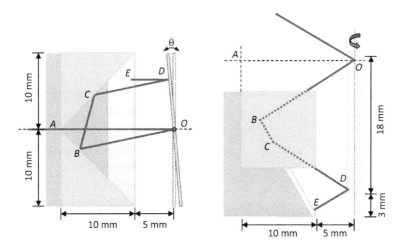

Figure 2 (a) Top view and (b) side view of the propagating path of reference beam in the optical delay line where the reference beam is incident at the pivot point of the scanning mirror.

By use of trigonometric relations, we can obtain the horizontal component of each segment of the light beam as

$$OB_{//} = \frac{1}{\cos 2\theta + \sin 2\theta} OA, \tag{1}$$

$$BC_{//} = \frac{2\sin 2\theta}{\cos^2 2\theta - \sin^2 2\theta} OA, \tag{2}$$

and

$$CD_{//} = \left[\frac{1}{\cos 2\theta + \sin 2\theta} - \frac{2\sin 2\theta \sin 3\theta}{\cos \theta (\cos^2 2\theta - \sin^2 2\theta)} \right] OA. \tag{3}$$

It can be proved that

$$OB_{//} + BC_{//} + CD_{//} = 2OA. \tag{4}$$

The real optical path length is then

$$OB + BC + CD = \frac{4}{\sqrt{3}} OA, \tag{5}$$

where a factor $2/\sqrt{3}$ is introduced because the beam is 30 degrees oblique in the vertical direction. The optical path length $OBCD$ is constant and is independent of the tilted angle θ of the scanning mirror. Therefore, the optical path difference of the optical delay line purely results from the variation of the length of DE. To calculate the length of DE, we need to know the relative position of the inclined reflection mirror to the scanning mirror. However, it is only the relative optical path difference that is important in our application. The optical path difference relative to the optical path length at $\theta = 0$ can be obtained as

$$\Delta L = 2\Delta DE = -(2\sqrt{3}\tan\theta\sin 3\theta)OA. \tag{6}$$

When the pivot of the scanning mirror deviates from the incident beam by δ as shown in Fig. 3, the optical delay line will result in a larger optical path difference. After calculation similar to the above, we obtain

$$AP + PB + BC + CD = 2\sqrt{3}(OA - \delta\tan\theta), \tag{7}$$

where θ is the tilted angle of the scanning mirror. The resulting optical delay due to the tilted angle of the scanning mirror θ is then

$$\Delta L_1 = -2\sqrt{3}\delta\tan\theta. \tag{8}$$

The calculation of the optical path difference in DE relative to the optical path length at $\theta = 0$ is more subtle and can be calculated to be

$$\Delta L_2 = -d\sin\left(\phi - \frac{\pi}{6}\right)\tan\theta, \tag{9}$$

where

$$d = \sqrt{d_1^2 + d_2^2}, \tag{10}$$

$$d_1 = \delta + 2\sin 2\theta(OA - \delta\tan\theta), \tag{11}$$

$$d_2 = \frac{\delta}{\sqrt{3}}, \tag{12}$$

$$\phi = \tan^{-1}\left(\frac{d_1}{d_2}\right). \tag{13}$$

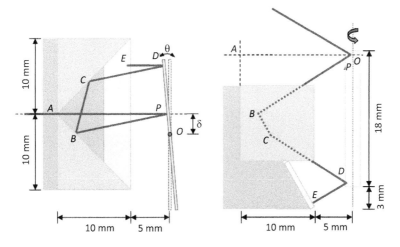

Figure 3 (a) Top view and (b) side view of the propagating path of reference beam in the optical delay line where the pivot of the scanning mirror is laterally deviated from the incident beam.

In our system, the scanning mirror was arranged 5 mm apart from the retro-reflector such that the length of *OA* was 15 mm. When the beam is incident at the pivot point of the scanning mirror, the allowable maximum tilted angle of the scanning mirror is 9.6 degrees and the achievable optical path difference, i.e., the scanning range, is 2.9 mm. When the pivot of the scanning mirror laterally deviates from the incident beam, the scanning range will be further improved as shown in Fig. 4.

Figure 5 shows the experimental setup of our low-coherence reflectometer with the use of the proposed optical delay line. In this system, a superluminescent diode (Superlum, D890-HP) with output power of 3.3 mW, center wavelength of 890 nm and spectral width of 150 nm was used as the light source. An He–Ne laser was used to calibrate the nonlinearity of the optical delay relative to the tilted angle of the scanning mirror. The galvoscanner was driven by a function generator which generated a continuous sawtooth waveform. The waveform was also used as a trigger to synchronize the acquisition of the interference signal received by the photodetector.

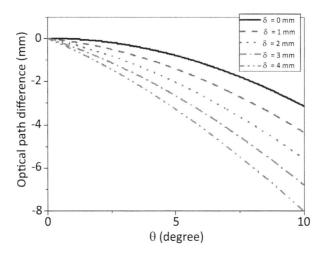

Figure 4 The relationship between the optical path difference (ΔL) and the tilted angle of the scanning mirror (θ) with different lateral deviation (δ) of the pivot of the scanning mirror from the incident beam. Beyond the allowable maximum tilted angle, the ray reflected from the retro-reflector will not able to be incident on the scanning mirror, except that a larger scanning mirror is used.

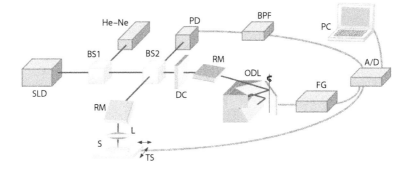

Figure 5 Schematic of the low-coherence reflectometer. SLD, superluminescent diode; He–Ne, He–Ne laser; BS1, beamsplitter (90/10); BS2, beamsplitter (50/50); RM, reflection mirror; L, lens; S, sample; TS, two-dimensional translation stage; DC, dispersion compensator; ODL, optical delay line; FG, function generator; PD, photodetector; BPF, band-pass filter; A/D, A/D converter; PC, personal computer.

The scanning rate of the optical delay line is determined by the scanning rate of the galvoscanner. In our experiment, a galvoscanner (GSI Lumonics, M3S) with 20 mm beam aperture mirror was used to perform a scanning rate of 200 Hz and the tilted angle was between 0 and 10 degrees. As shown in Fig. 6, there were two A-scans performed in opposite directions in each cycle of scanning. Therefore, the scanning rate of the low-coherence reflectometer in our experiment was doubled to be 400 Hz. The scanning rate of the optical delay line can be improved without changing the configuration of the system when a higher scanning rate of the galvoscanner is applied.

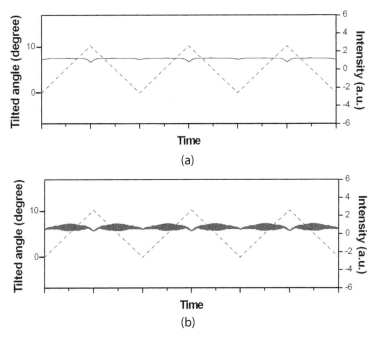

Figure 6 (a) The intensity of the reflected beam from the reference arm of the low-coherence reflectometer during scanning. (b) The interference signal when an He–Ne laser was used as the light source.

Design and Construction: Example 2

Optical delay line is composed of a retro-reflector, an inclined reflection mirror, and a scanning mirror, as shown in Fig. 1(a). The retro-reflector is fabricated with two right-angled prisms assembled on an aluminum jig where a reflection mirror inclined at 60 degrees is integrated. Fig. 1(b) shows the photograph of the retro-reflector assembled with the reflection mirror where one can see that the size of the entire system is within 2 cm × 2 cm. As shown in Fig. 1(a), the light beam is incident at 30 degrees downward on the scanning mirror. After reflecting by the scanning mirror and the retro-reflector, the beam is reflected by the scanning mirror again. Finally, the beam is normally incident on the inclined reflection mirror such that the reflected beam is collinear with the incident beam. Since the reflected beam is guaranteed to be collinear with the incident beam regardless of the tilted angle of the scanning mirror, no walk-off problem will be generated in such an optical delay line. Further detailed design was mentioned somewhere else [20].

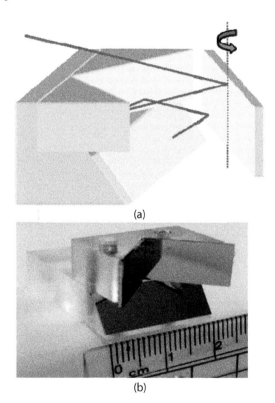

(a)

(b)

Figure 1 (a) Design of optical delay line. (b) Photograph of retro-reflector assembled with reflection mirror.

In Fig. 2 we show the experimental setup of the low-coherence reflectometer with the use of the proposed optical delay line. In our system, a superluminescent diode (Superlum, D890-HP) with output power of 3.3 mW, center wavelength of 890 nm, and spectral width of 150 nm was used as the light source. An He–Ne laser was used to calibrate the nonlinearity of the optical delay relative to the tilted angle of the scanning mirror. The galvoscanner was driven by a function generator which generated a continuous sawtooth waveform. The waveform was also used as a trigger to synchronize the acquisition of the interference signal received by the photodetector. Scanning rate of the optical delay line is determined by the scanning rate of the galvoscanner. In this experiment, a galvoscanner (GSI Lumonics, M3S) with 20 mm beam aperture mirror was used to perform a scanning rate of 200 Hz and the tilted angle was between 0 and 10 degrees.

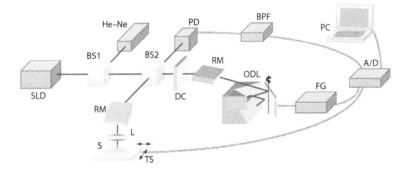

Figure 2 Schematic diagram of low-coherence reflectometer. SLD, superluminescent diode; He–Ne, He–Ne laser; BS1, beamsplitter (90/10); BS2, beamsplitter (50/50); RM, reflection mirror; L, lens; S, sample; TS, two dimensional translation stage; DC, dispersion compensator; ODL, optical delay line; FG, function generator; PD, photodetector; BPF, band-pass filter; A/D, A/D converter.

As shown in Fig. 3, there were two A-scans performed in opposite directions in each cycle of scanning. Thus, the scanning rate of the low-coherence reflectometer in the experiment was doubled to be 400 Hz. The scanning rate of the optical delay line can be improved without changing the configuration of the system when a higher scanning rate of the galvoscanner is applied.

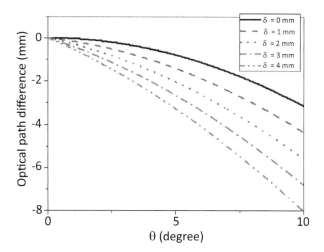

Figure 3 Optical path difference (ΔL) with respect to the tilted angle of the scanning mirror (θ) with different lateral deviation (δ) of the pivot of the scanning mirror from the incident beam.

Design and Construction: Example 3

In Fig. 1 we show the compact optical delay line for free-space optical coherence tomography configuration, including the retro-reflector, the inclined reflection mirror, and the scanning mirror. The retro-reflector is constructed by using the right-angled prisms pair mounted onto the aluminum jig. The reflection mirror is tilted at 60° normal to the surface of the scanning mirror and it is located below the prism pair under the same aluminum jig. The gap between the retro-reflector and the scanning mirror at its normal surface is 5 mm. The total dimension of our delay line is about 2 cm × 2 cm.

The experimental setup of the compact optical delay line in the typical free-space optical coherence tomography configuration is described as in [24]. Briefly, the He–Ne laser is used as the light source and it is also performed to calibrate the nonlinearity of the optical delay line with respect to the tilted angle of the scanning mirror.

The tilted angle is set between 0° and 10°. First, the light is incident at 30° normal to the scanning mirror's surface. After light reflection in ordering at the scanning mirror and the retro-reflector, the light is incident at the scanning mirror once again. The light is then incident at the inclined reflection mirror so that the reflected light is firmly collinear with the incident one. This can avoid the walk off problem causing variation in intensity of optical signal during scanning. It is noted that the galvoscanner is driven by a function generator to generate continuously waveform. This waveform is served as a trigger to synchronize the acquisition of the interference signal. The scanning rate of the delay line is determined by the galvoscanner's scanning rate at 400 Hz.

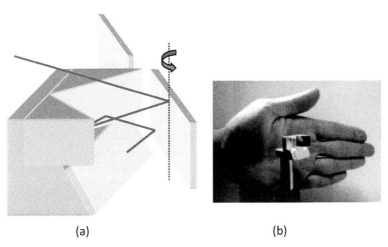

(a)   (b)

Figure 1   Conceptual design (a) and photograph (b) of the compact optical delay line for the optical coherence tomography including the retro-reflector integrated by the inclined reflection mirror. The red line represents the light propagation in the delay line system indicating the direction of the light from/to the delay line is in collinear conditions during scanning.

5.3   Quantitative and Qualitative Methods

Basic and applied researches can be quantitative or qualitative or even both.

1. Quantitative data are based on the measurement of quantity. A process is described in terms of one or more quantities. The data of this research are a set of numbers by using statistical or mathematical analysis. The results are often presented in tables and graphs.

2. Qualitative data are concerned with quality. Qualitative methods can be used to understand the meaning of the numbers that we obtain from quantitative research. The data of this research cannot be graphed. Experimental and simulation studies are quantitative research.

5.4 Problems

Pb.1. Discuss the difference between experiment method and design and construction method.

Pb.2. Indicate the details needed for experiment method.

Pb.3. Indicate the details needed for design and construction.

Pb.4. If you are dealing with theoretical simulation, write how the experiment looks like.

Pb.5. Comment on the following experiment methods:

Experiment (A)

At the first step of the experiment, Cu nanowires of 19 nm width × 40 nm periodicity on NaCl(110) template were fabricated by a shadow deposition method in vacuum [2]. The variation in SH response was then measured when the sample was rotated azimuthally in the surface plane at an excitation photon energy of 1.17eV driven by a ps-pulsed Nd:YAG laser [4]. The experimental results were compared with the theoretical work by finite-difference time-domain (FDTD) method to gain insight into the SH process from these samples.

Experiment (B)

NaCl(110) single crystals were mounted in ultrahigh vacuum deposition chamber. Layers of NaCl were deposited after annealing substrate. The SiO layers were then deposited. Desired metal was deposited at a given deposition rate onto the substrate at the flux angle of 65° with respect to the surface normal to form nanowires. On arrays of metallic nanowires the SiO layers were deposited. For transmission electron microscopy (TEM) observation, metallic arrays sandwiched by SiO layers were separated from NaCl substrates by dissolving substrates in distilled water, and mounted on Cu grids.

Light source of the fundamental photon energy from 1.2 to 2.3 eV was an optical parametric generator and amplifier system driven by a mode-locked Nd:YAG laser. The excitation beam was focused on the sample surface at the incident angle of 45°. The SH signal from sample was then detected by a photomultiplier. The measurements were carried out in p-in/p-out polarization combination.

Experiment (C)

Sample Preparation. NaCl(110) single crystals were mounted in an ultrahigh vacuum deposition chamber. The layers of NaCl were deposited after annealing the substrate. The SiO layers were then deposited. Desired metal was deposited at the given deposition rate onto the substrate at the flux angle of 65° with respect to the surface normal to nanowires. On the arrays of metallic nanowires the SiO layers were deposited once again. For transmission electron microscopy (TEM) observation, metallic arrays sandwiched by SiO layers were separated from the NaCl substrates by dissolving the substrates in distilled water, and mounted on 200–400 mesh Cu grids.

Azimuthal Angle Dependence of the SH Intensity. Probe-light pulses with a fundamental photon energy of 1.17 eV were generated by a mode-locked Nd:YAG laser. The fundamental light beam was focused onto the sample surface and the incident angle was 45°. The SH signal was detected by a photomultiplier and the absolute SH intensities were obtained by normalizing the SH response to that of the α-SiO$_2$ (0001). To measure the azimuthal angle dependence of the SH intensity, the samples were mounted on an automatic rotation stage with the surface normal set parallel to the rotation axis of the stage. The measurements were carried out in four different input/output polarization combinations: p-in/p-out, p-in/s-out, s-in/p-out, and s-in/s-out.

Wavelength Dependence of the SH Intensity. Light source of the fundamental photon energy from 1.2 to 2.3 eV was an optical parametric generator and amplifier system driven by a mode-locked Nd:YAG laser. The excitation beam was focused on the sample surface at the incident angle of 45°. The SH signal from the sample was detected by a photomultiplier. Finally, the absolute magnitudes were obtained by normalizing the SH response to that of the α-SiO$_2$ (0001). The measurements were carried out in p-in/p-out polarization combination.

Experiment (D)

NaCl(110) single crystal was mounted in a ultrahigh vacuum deposition chamber. The layers of NaCl were deposited after annealing the substrate. The SiO layers were then deposited. Cu was deposited at the rate of 0.6 nm/min onto the substrate at the flux angle of 65° with respect to the surface normal to form nanowires or nanodots. On the arrays of Cu nanowires and nanodots, the SiO layers were deposited once again. For transmission electron microscopy (TEM) observation, Cu nanowire arrays sandwiched by SiO layers were separated from the NaCl substrates by dissolving the substrates in distilled water, and mounted on 200 mesh Cu grids. On the other hand, absorption spectra at the normal incidence were measured as a function of the sample rotation angle. These spectra were recorded in the wavelength 250–1100 nm by a spectrometer equipped with a polarizer accessory, a Xe lamp source, and a photomultiplier tube.

Experiment (E)

In this study, the incident photon wave at excitation wavelength of 760 nm diffuses from a point source, as shown in Fig. 1.

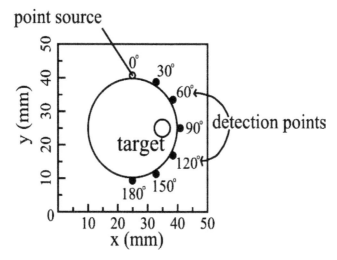

Figure 1   Coordinate system in FDOT measurement.

The source was a pico-second pulsed laser. A 2 µM ICG in 1% Intralipid solution served as the fluorescent target filled in a cylindrical polyoxymethylene (POM) phantom, with 60 mm in length and 30 mm in diameter, was excited by the incident wave and behaves as secondary point sources of fluorescent light propagating at the emission wavelength of about 850 nm toward the appropriated filtered detection system. The fluorescence temporal profiles were recorded at detection angles ranging from 30° to 180°, by 30° steps, from the 0° irradiation point as shown in Fig. 1. In Fig. 1, an ICG target was 2 mm, 4 mm, or 6 mm in diameter and was located 11.5 mm, 11.5 mm, or 10.5 mm, respectively, away from the center of the cylindrical phantom. The target was excited at the perpendicular position (00) from the target symmetrical plane as shown in Fig. 1. Then, the target was clearly observed on the symmetrical geometry between 30°–150° and 60°–120° detection points.

Experiment (F)

The second harmonic intensity from the Au, Cu, and Pt nanowire arrays was measured as a function of the sample rotation angle φ at the photon energy of 1.17 eV for Au and Cu and at the photon energy of 2.33 eV for Pt. The angle ϕ was defined as the angle between the incident plane and the wire axes in the [001] substrate direction. The measurements were carried out in four different input/output polarization combinations: p-in/p-out, p-in/s-out, s-in/p-out, and s-in/s-out. The sample preparation method was shadow deposition and outlined as in [9]. The minimum widths of the obtained nanowires were 40 nm for Au, 14 nm for Cu, and 9 nm for Pt. By using this shadow deposition technique, we obtained the large areas of very thin samples and we could perform the transmission electron microscopy and optical experiments at different points in each sample as well as compare the obtained data statistically. Furthermore, the ridge-and-valley surface morphology was considered to be preserved well even after the deposition of amorphous SiO overlayer, as confirmed for amorphous carbon overlayer by the AFM study [10].

Pb.6. Comment on the following design and construction methods:

Design and Construction (A)

We have numerically investigated the local field enhancement at the fundamental frequency related to the χ^2_{311} and χ^2_{323} elements from Cu nanowires by the FDTD method. Commercial software Lumerical

FDTD solutions 4.0 was used for our calculation [4]. Three arrays of Cu nanowires are defined to be infinitely extended in direction 1. Since the second and third subscripts of the nonlinear susceptibility elements represent the directions of the fundamental field at frequency ω, two different polarizations of incident light were considered: Polarizations A and B denote the electric field polarized in (2, 3) and 1 directions and are involved in the nonlinear optical process represented by the χ^2_{323} and χ^2_{311} elements, respectively. In addition to the circular wires, the square and triangular wires have the same cross-sectional areas as the circular ones so that the field distributions should be comparable. The diameter of the circular wires was 20 nm. The perimeters of the square and triangular wires were 72 and 84 nm, respectively. The periodicity of the wires was changed between 4 and 40 nm. We used Drude–Lorentz parameters fit to the empirical dielectric constant data of copper obtained from Ref. [7] in the photon energy of interest. The FDTD calculation was performed for the mesh size of 1 nm × 1 nm. A typical propagation was over a time period of 500 fs. The total field intensity was normalized to that of the intensity of the incident field. For all simulations, absorbing boundary conditions based on the perfectly matched layer approximation were applied [8].

Design and Construction (B)

The sum frequency (SF) intensity images of ZnS polycrystals were obtained using the setup shown in Fig. 1. The ZnS pellet grown by chemical vapor deposition (CVD) with 10 mm diameter and 3 mm thickness was purchased from Furuuchi Chemical Co. Ltd, Japan. The visible light at λ_{VIS} = 532 nm was the doubled-frequency output from a mode-locked Nd:YAG laser operating at the repetition rate of 10 Hz with the pulse duration time of 35 ps. The wavelength tunable IR light was an output from an optical parametric generator and amplifier system driven by the fundamental and second harmonic output of the same Nd:YAG laser. The pulse energies of the visible and IR beams at the sample surface were 2.5 and 5 μJ/pulse, respectively. The visible and the IR beams were incident on the sample with the incident angles of 0° and 50°, respectively.

The SF intensity was measured at an IR wave number of 2890 cm^{-1} (λ_{IR} ~ 3 μm). The SF signal appeared at λ_{SF} ~ 460 nm. The SF light from the sample was collected by an objective lens of the microscope optics. The imaging optics was a commercial microscope (Nikon Eclipse: LV100D). The SF light was then reflected by a dichroic mirror, passed through a bandpass filter, and was focused into a pinhole of 2 mm or 0.4 mm diameter. We used no input aperture because the input visible beam at λ_{VIS} was an ideal parallel beam. The magnification of the optics was ×20 at the output pinhole. Namely, the image at the sample was expanded by 20 times at the pinhole.

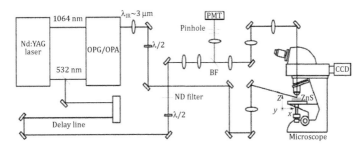

Figure 1 Experimental setup for the measurements of sum frequency (SF) intensity images using the optical confocal SF microscopy. OPG/OPA represents the optical parametric generator and amplifier. PMT represents a photomultiplier. BF represents the birefringent filter. CCD camera represents the charge-coupled device camera.

The SF signal was finally detected by a photomultiplier (PMT) and the absolute SF intensities were obtained by normalizing the SF response to that of the incident light. The size of the scanned area was 100 × 100 μm^2 with the step size of 0.5 μm. The measurement time for one image was 60 to 90 minutes. Linear image was monitored by a charge-coupled device (CCD) camera. All SFG experiments were performed in air at room temperature.

Chapter 6

Data Analysis of Report

In this chapter, I will show how to perform data analysis to obtain accurate results so that readers can replicate them. The third major section of a research paper includes results and discussion. The results section contains facts related to your experiment. You should use figures, tables, and equations so that understanding your data becomes easier. The results are interpreted in the discussion section.

6.1 Results and Discussion

After performing the experiment on your chosen problem, you should prepare the section on results and discussion. In this section, you should give all evidence relevant to the research problem and its solution. A bare statement of the findings is not enough; the implications need to be informed. Data analysis and the report must be accurate. These must be based on the research questions you have formulated. This section must include statistical operations. You compare your findings to see whether they agree with previous research. You inform the strengths and weaknesses of your work and your suggestions for the study. Also you can show the direction of your future research.

Research Methodologies for Beginners
Kitsakorn Locharoenrat
Copyright © 2017 Pan Stanford Publishing Pte. Ltd.
ISBN 978-981-4745-39-0 (Hardcover), 978-1-315-36456-8 (eBook)
www.panstanford.com

The organization of the results section should be as follows:

1. You start with a paragraph, not tables or figures.
2. You produce tables and figures after mentioning them in the text.
3. You can explain if any data are missing or problems exist.
4. You explain the main results and compare your expectation (hypothesis) with that of other researchers.
5. You explain all other interesting trends in your data.

The results and discussion section can be revised anytime based on the following:

1. Quantitative and qualitative analyses are adequately performed to draw the conclusion.
2. The results and discussion are general.
3. The results and discussion are valid for the situation considered in the present work.
4. The discussion is not too broad considering the analysis performed.
5. The evidence is not too weak for the discussion.

6.1.1 Table Format

Tables are good for showing exact values or a lot of different information. Tables (refer them with Table 1, Table 2, ...) should be presented as part of the text, but avoiding confusion with the text. A descriptive title should be placed above each table. Units in tables should be given in square brackets [meV]. If square brackets are not available, use curly {meV} or standard brackets (meV). Special signs, for instance α, γ, μ, Ω, (), $\geq$, $\pm$, •, Γ, $\{11\bar{2}0\}$, are always written in the fonts Times New Roman or Arial. Here is an example of the table format.

6.1.1.1 Example of table

Table 1 Geometrical and deposition parameters of the fabricated samples

Sample	Wire width (nm)	Nominal thickness (nm)	Periodicity (nm)	Homoepitaxial growth temperature (°C)
C7	17	20	27	200
C8	14	10	27	200
C9	19	20	40	400
C10	15	10	40	400
C12	$r^* = 5$	5	22	200

r^* is the radius of nanodots.

6.1.2 Figure Format

Graphs are good for showing the overall trends and are much easier to understand. Graphs and tables should not depict the same data. Figures (refer them with Fig. 6.1, Fig. 6.2, …) should be presented as part of the text, leaving enough space so that the caption is not confused with the text. The caption should be self-contained and placed below or beside the figure. Generally, only original drawings or photographic reproductions are acceptable. Utmost care must be taken to correctly align the figures with the text. If possible, you should include your figures as images in the electronic version.

For the best image quality, the pictures should have a resolution of 300 dpi (dots per inch). Color figures are good for the online version of the journal. Generally, these figures will be produced in black and white in the print version, partly for reducing the cost of publication. You should indicate on the checklist of the publication if you wish to have the figures printed in full color and make the necessary payments in advance. You can use many different types of graphs to show your results: line graph, scatter plot, bar graph, histogram, etc.

6.1.2.1 Examples of figures

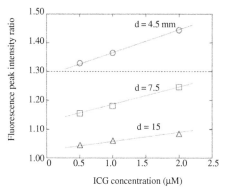

Figure 1 Indocyanine green (ICG) concentration as a function of fluorescence peak intensity ratio at a detection point of 30°/–30°. The target was 6 mm in diameter and was located at d = 4.5 mm, 7.5 mm, or 15 mm, measured from the phantom surface to the target center.

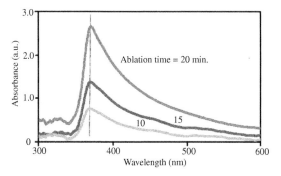

Figure 2 Measured absorption spectra of the zinc oxide nanoparticles with various ablation times.

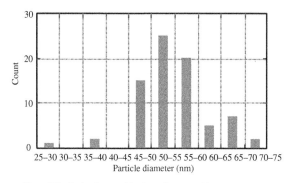

Figure 3 Distribution diagram of the zinc oxide nanoparticles.

6.1.3 Equation Format

Equations (refer them with Eq. 1, Eq. 2, . . .) should be indented 5 mm (0.2"). There should be one line of space above and below the equations. They should be numbered sequentially, and the number should be put in parentheses at the right-hand margin. Equations should be punctuated as if they were part of the text. Punctuation appears after the equation but before the equation number, such as

$$c^2 = a^2 + b^2. \tag{1}$$

6.1.4 Examples of Results and Discussion

Results and Discussion: Example 1

Results

TEM images in Fig. 1 show the uniform Cu nanowires as long and straight periodic dark images parallel to the [001] direction of the substrates. The nominal thicknesses of the Cu nanowires were measured with the sensor plane of the thickness monitor perpendicular to the Cu beam direction and are shown in Table 1. The ridge-and-valley surface morphology is considered to be preserved well even after the deposition of amorphous SiO overlayer, as confirmed for amorphous carbon overlayer by AFM study [11]. At the nominal thickness of 5 nm, Cu forms arrays of nanodots (Figs. 1e and f). The arrays of nanowires are formed at the nominal thickness of 10 nm, and the minimum widths of 14 nm are obtained (Fig. 1b).

Cu nanodots fabricated in the present work have spherical shapes as seen in Fig. 1f and thus the polarization dependence of the plasmon resonance at 2.25 eV is not observable as is seen in Fig. 2. These absorption spectra are in agreement with the strong absorption band at about 2.21 eV of spherical Cu nanoparticles in the literatures [12,13]. This result for nanodots serves as a control experiment for the results for nanowires in Fig. 3. In the absorption spectra of Cu nanowires, we observe absorption maxima near 2.25 eV for perpendicular polarization in Fig. 3b and they are redshifted for the parallel polarization in Fig. 3a. We also find the red-shift of the absorption maxima when the widths of nanowires are increased. The nanowires of smaller widths have more anisotropic absorption than those of larger widths. Variety of the periodicity of the wires through the NaCl homoepitaxial growth

Compare the findings with other studies.

temperature does not give rise to significant changes in these spectra. Our measurements of the energy shifts of the absorption maxima were quite reproducible among several runs at different positions from each sample within the coefficient of variation of less than 5%.

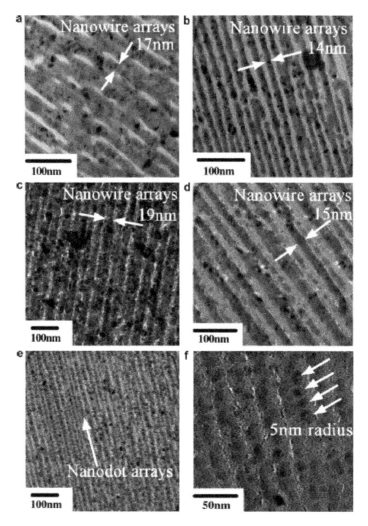

Figure 1   TEM images of Cu nanowires of samples: (a) C7, (b) C8, (c) C9, and (d) C10, and TEM images of Cu nanodots of sample C12 at (e) low magnification and (f) high magnification. The deposition parameters are shown in Table 1.

Table 1 Geometrical and deposition parameters of the fabricated samples

Sample	Wire width (nm)	Nominal thickness (nm)	Periodicity (nm)	Homoepitaxial growth temperature (°C)
C7	17	20	27	200
C8	14	10	27	200
C9	19	20	40	400
C10	15	10	40	400
C12	$r^* = 5$	5	22	200

r^* is the radius of nanodots.

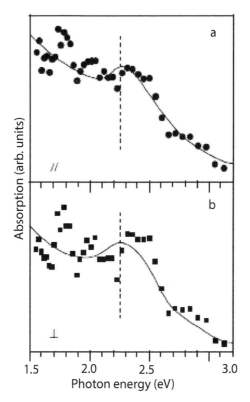

Figure 2 Absorption spectra of 10 nm Cu nanodots on the faceted NaCl(110) template. // (⊥) denotes the electric field parallel (perpendicular) to the wire axes. Solid line is guide to the eyes.

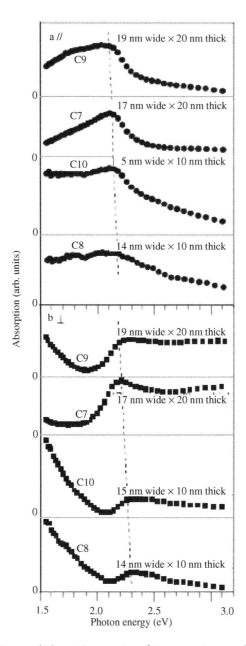

Figure 3 Measured absorption spectra of Cu nanowires sandwiched by SiO layers. // (⊥) denotes the electric field parallel (perpendicular) to the wire axes.

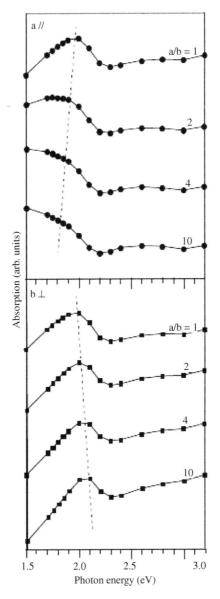

Figure 4 Calculated absorption spectra of Cu nanowires at different aspect ratio *a/b*. *a* (*b*) is the long (short) wire axes. // (⊥) denotes the electric field parallel (perpendicular) to the wire axes.

Discussion

One of the key parameters determining the minimum width of the nanowires deposited onto a crystalline substrate is the surface energy of the wire metal. High surface energy of metals will lead to high contact angle on the substrate surface. As the surface energy is the largest for Pt, the second largest for Cu, and the smallest for Au in their liquid state [14], the minimum width of Cu nanowires (14 nm) is larger than that of Pt (9 nm) [15], but is smaller than that of Au (40 nm) [8].

Figure 3 shows that blue-shift of the absorption maxima occurs with the small wire widths. Since the wire dimensions and separation distances are relatively small compared to the wavelength of the incident light, we have applied Maxwell–Garnett theory to address this size effect on the absorption spectra [16]. This theory defines an effective dielectric constant for a composite from the dielectric constants of constituent metal and the host medium. By assuming the metal volume fraction of 0.1 and various aspect ratios a/b (a and b are defined as the long and short wire axis lengths, respectively), we calculate their absorption spectra as shown in Fig. 4. The absorption maxima for the incident field polarized along the wire axes occur at lower photon energy than those for the incident field polarized perpendicular to the wire axes. This is consistent with the experimental result, but we find that the detailed experimental dependence of the energy positions of the absorption maxima on the polarization is different from the prediction by Maxwell–Garnett model.

Compare the findings with other studies.

There are four candidate origins of the discrepancy of the peak energy positions between the experiment and the theory. First, the dielectric constants of the composite material may not be simply the weighed averages of those of the constituents. Second, the particle size at the long particle axis might be larger than the optical wavelength in the visible region so that the non-retardation limit is not applicable [16]. Third, there may be changes of the electron density profile at the particle surface due to the wire–matrix interaction [13]. Finally, the collision process at the wire walls may result in a reduced mean free path of the conduction electrons [17].

Give your own suggestions

Since the Maxwell–Garnett model cannot reproduce the systematic changes in the polarization dependences of absorption intensity qualitatively, we suggest that this model is not appropriate for predicting our

Give your own suggestions

absorption spectra. In order to remove the first and second sources of discrepancies mentioned above, rigorous numerical studies using the finite-difference time-domain method of solving the electromagnetic problems by integrating Maxwell's differential equations are under way. This method will also allow us to consider the actual wire size dependence of the peak energy positions.

Results and Discussion: Example 2

Figure 1 shows the measured absorption spectra of Cu nanowire arrays with various widths and nominal thicknesses. The nominal thicknesses were measured with the sensor plane of the thickness monitor set perpendicular to the Cu beam direction. Plasmon resonance structures are centered around 2.3 eV when the incident field is perpendicular to the wire axes as in Fig. 1(a) and they shift to lower energy when the field is parallel to the wire axes as in Fig. 1(b). In both polarization configurations, they shift to lower photon energy when the widths of the nanowires are increased.

In the case of metal spheres embedded in dielectric materials, Mie scattering theory has succeeded in describing extinction spectra in the visible region, including the surface plasmon resonance of dipole modes [4]. For the nanowires in a nonabsorbing medium, the plasmon resonant modes are described by Maxwell–Garnett model by introducing a shape-dependent screening factor k for different dimensions [9].

Screening factor k is related to the depolarization factor q via the relation $k = q^{-1} - 1$. If metal nanowires are regarded as the ellipsoidal elongated particles, $q_{||}$ $(q_{\perp})$ for an electric field incident along the long axis a (the short axis b) is estimated via the relation $q_{||} \approx [1 + (2a/b)]^{-1}$ $(q_{\perp} \approx [2 + (b/a)]^{-1})$. Surface plasmon resonance results in enhancement of absorption coefficients (Abs) at the resonance frequency, and it can be understood in terms of the particle polarizability α as

Abs = $k \, \mathrm{Im}\{\alpha\}$

$$\alpha = \frac{(V/3q)(\varepsilon_m - \varepsilon_0)}{(\varepsilon_m + \kappa\varepsilon_0)},$$ (1)

where k is the wave vector ($= 2\pi/\lambda$) and V is the particle volume. ε_m and ε_0 are the complex dielectric functions of the metal and the surrounding medium, respectively. Figure 2 shows absorption spectra

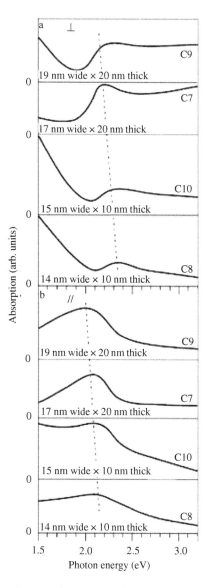

Figure 1 Measured absorption spectra of Cu nanowires. C7–C10 indicate the sample name listed in each figure. ⊥ (//) means the electric field perpendicular (parallel) to the wire axes.

calculated with Maxwell–Garnett model for metallic ellipsoidal particles with different aspect ratios. Plasmon maxima appear near

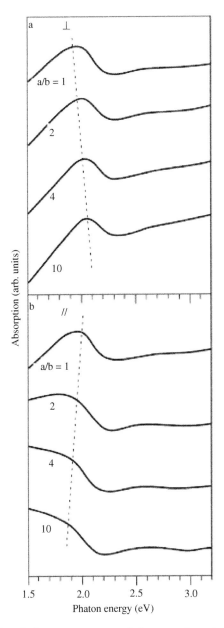

Figure 2 Calculated absorption spectra of Cu nanowires at different aspect ratios *a/b* by Maxwell–Garnett model. *a* (*b*) is the long (short) wire axes of the nanowires. ⊥ (//) represents the electric field perpendicular (parallel) to the wire axes.

2.0 eV for the electric field perpendicular to the nanowire axes in Fig. 2(a), and they shift to higher photon energy with increasing aspect ratio. On the other hand, plasmon maxima are observed below 2.0 eV for the electric field parallel to the wire axes in Fig. 2(b), and they shift to lower energies with increasing aspect ratio. The resonant condition between the dielectric constant of Cu and the surrounding medium is explained as $Re\{\varepsilon_m\} = -Re\{\kappa\varepsilon_0\}$ [10]. With the increase of the aspect ratio, the screening factor k of plasmon resonance for the electric field parallel to the wire axis approaches infinity and the resonance peak shifts toward lower photon energy accordingly.

However, Maxwell–Garnett model with Eq. (1) attempts only to address the aspect ratio dependence of the electric dipole response and not the dependence on the absolute sizes of the nanostructures. The actual sizes of our nanowires are over the limitation of the dipole approximation applicable to sizes much smaller than the wavelength of light. Thus, we attempt a prediction by more rigorous FDTD method to analyze the absorption spectra. This model will allow us to consider the size dependence of the electric dipole resonance.

In FDTD simulation as shown in Fig. 3, nanowires are surrounded by absorbing boundaries with small meshes of 0.1 nm periodicity, yielding geometry parameters for the long wire axes of $a = 140$–190 nm and the short axes of $b = 14$–19 nm. Simulation time is 200–500 fs to capture all over the photon energy of 1.5–3 eV. Simulation is performed until there are essentially no electromagnetic fields left in the simulation region. If we drop one of the spatial dimensions, for example, z-direction, and consider the x–y plane, the 2-D Maxwell's equations split into two independent sets of equations consisting of three vector quantities each. These are termed the TE (transverse electric) and TM (transverse magnetic) equations [11]. By considering x-polarized light traveling in the forward y-direction and x-polarized light traveling in the forward y-direction, the three electromagnetic field components Ex(x,y,t), Ey(x,y,t), and Hz(x,y,t) for each polarization direction are calculated. Optical properties of metals can be well described by using Drude model supposing that zero restoring force is applied to free electrons. Then, the dielectric function of metals is written as:

$$\varepsilon_m \approx \varepsilon_0 \left(1 - \frac{\omega_p^2}{\omega^2} \right). \tag{2}$$

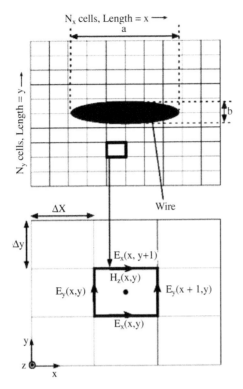

Figure 3 Schematic image of a sample system and coordinate used in the two-dimensional FDTD calculation. a (b) is the long (short) wire axes of the ellipsoid.

We have used the dielectric function defined in the material database of a commercial software Lumerical FDTD Solutions 4.0 [12]. The built-in material database contains dielectric properties of copper obtained by fitting the Drude dielectric function to measured optical constants of copper in the frequency region of interest [13]. By this FDTD method, we have calculated absorption spectra as shown in Fig. 4 as a function of the photon energy for the ellipsoidal nanowires, and with the wire widths as the minor axis lengths of $b = 14$–19 nm at a constant aspect ratio of $a/b = 10$. This simulation by FDTD calculation is more favorable than that of Maxwell–Garnett model from a point of view of the reproduction of the experimental data. Calculation results show the plasmon maxima near 2.4 eV when the electric field is perpendicular to the wire axes as shown in Fig. 4(a), while the red-shift occurs when the electric field is parallel to the wire axes as shown in Fig. 4(b). The peak positions in the latter polarization configuration are above 2.0 eV, as is consistent with the experimental results in Fig. 1(b). Increasing the wire widths by increasing the size of minor axis of the ellipse (*b*) also induces a red-shift of the peak positions.

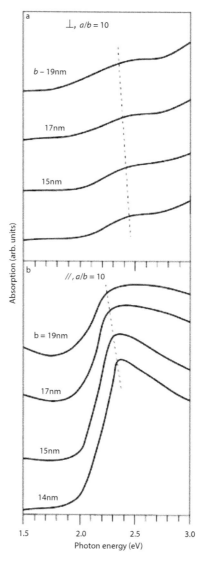

Figure 4 Calculated absorption spectra of Cu ellipsoids at a fixed aspect ratio of *a/b* = 10 by FDTD method. *a* (*b*) is the long (short) axes of the ellipsoids. ⊥ (//) shows the electric field perpendicular (parallel) to the longer axes.

The disagreement between the experiment and the calculation by Maxwell–Garnett model in terms of the polarization dependence of the absorption spectra in Fig. 2 suggests that the calculation by this model becomes inapplicable when the wire size is not negligible compared with the wavelength of incident light. When the electric field is perpendicular to the wire axes as shown in Fig. 2(a), the calculated positions of plasmon maxima are observed near 2.0 eV in contrast to the experiment in Fig. 1(a) showing plasmon maxima around 2.3 eV. When the electric field is parallel to the wire axes as shown in Fig. 2(b), plasmon maximum energies are observed below 2.0 eV, and they conflict with the experiment in Fig. 1(b) showing plasmon maxima above 2.0 eV. Sandrock et al. [10] have reported that the small particle limit treatment of such Maxwell–Garnett model is qualitatively correct in local field enhancement calculation only for systems of small short particle axes (b becomes 0). However, when the retardation effect of the incident light fields was taken into account, calculation gave different results from those predicted by the quasi-static theory for the systems of $b = 15$ nm. Therefore, we can say that the nonretardation limit may not be applicable to our study with short wire axes of $b = 14$–19 nm. The present calculation taking into account the absolute wire size by FDTD simulation is judged to have improved the agreement between the experiment and the theory because it takes full account of the retardation effect.

> Compare the findings with other study

There still remains a discrepancy between the theory and the experimental data. The measured spectra in Fig. 1(a) of samples C9, C10, and C8 show increase toward the lower photon energy, whereas the one in Fig. 4(a) calculated by FDTD method displays monotonical reduction on the lower photon energy side. Using the Drude model alone for our study is probably the reason for this discrepancy. In the case of gold, Krug et al. [14] have determined the optical constants of Au nanoparticles with a modified Debye function for the FDTD method, and Vial et al. [15] have confirmed that the Drude–Lorentz models could fit the dispersion curve of Au nanostructures over a range wider than the single Drude model. Since the Drude model in Eq. (2) may not provide an accurate model of empirical dielectric constant data for copper over a wide frequency range, a combination of at least one more dispersion model dielectric functions is suggested to be added in order to fit the dielectric function of copper over the full optical spectrum. For further development of the calculation,

> Compare the findings with other study

> Give your own suggestions

it is necessary to consider the contributions from the confinement effect and the collision process at the wire walls of Cu electrons [7].

Results and Discussion: Example 3

Results

Figures 1 and 2 display the azimuthal angle dependence of the SH intensity measured in air from the Cu nanowire arrays of different wire widths and from the Cu nanodots of sizes around 10 nm. The SH intensity is measured as a function of the sample rotation angle ϕ at the photon energy of 1.17 eV. The angle ϕ is defined as the angle between the incident plane and the wire axes in the [001] direction. In each row the SH intensity patterns for four different combinations of input and output polarizations are shown for each sample. The intensity is plotted in a relative scale shown in each pattern.

The SH intensity from Cu nanowires in Fig. 1 depends strongly on the sample rotation angle ϕ and the polarization combinations of the fundamental and SH light. We see two main lobes in p-in/p-out, s-in/p-out, and s-in/s-out polarization configurations in the first, third, and fourth columns, respectively. The SH intensity is as low as the noise level for p-in/s-out polarization configuration in the second column. The SH intensity from Cu nanodots in Fig. 2 depends weakly on the sample rotation angle ϕ and the polarization configuration. We see the isotropic structures in p-in/p-out and s-in/p-out polarization configurations in Figs. 2(a) and (c), respectively. The SH intensity for p-in/s-out in Fig. 2(b) and for s-in/s-out polarization configurations in Fig. 2(d) is below the noise level.

Discussion

The SH response from 10 nm Cu nanodots in Fig. 2 is a clear indicator of their isotropy in nonlinear response in any polarization combinations. This indicates that the Cu–SiO–NaCl interfaces do not contribute to the anisotropies of SH responses. The observed anisotropy in the nonlinear optical susceptibility elements from Cu nanowires is therefore considered to originate from their anisotropic wire shapes.

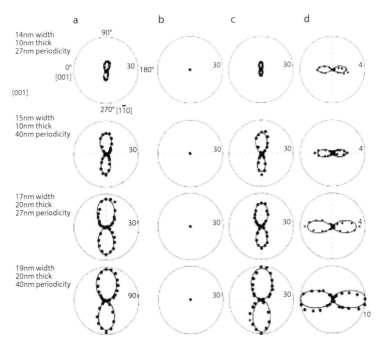

Figure 1 Measured (filled circles) and calculated (solid line) SH intensity patterns for different wire widths of Cu nanowires as a function of the sample rotation angle ϕ at the fundamental photon energy of 1.17 eV. The angle ϕ is defined as the angle between the incident plane and the [001] direction on the sample surface. The incident angle is 45°. The SH intensity is plotted in polar coordinates. The intensity in a relative unit is written next to each pattern. The four different input and output polarization combinations are indicated at the top: (a) p-in/p-out, (b) p-in/s-out, (c) s-in/p-out, and (d) s-in/s-out.

We have analyzed the SH intensity patterns from Cu nanowires in Fig. 1 (filled circles) by a phenomenological model [5]. We have first fitted the theoretical SH intensity patterns to the experiment by a least square fitting program assuming C2v symmetry for the Cu nanowires/ NaCl(110) system. This symmetry permits the five independent nonlinear susceptibility elements: $\chi^2_{113}, \chi^2_{223}, \chi^2_{311}, \chi^2_{322}, \chi^2_{333}$. The indices 1, 2, and 3 denote the [001], [1$\overline{1}$0], and [110] directions on the NaCl(110) substrates, respectively. However, the calculated patterns could not reproduce the SH intensity in s-in/s-out polarization

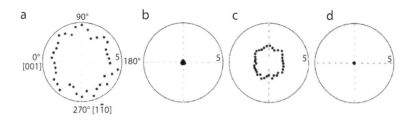

Figure 2   Measured SH intensity patterns for 10 nm Cu nanodots as a function of the sample rotation angle ϕ at the fundamental photon energy of 1.17 eV. The angle ϕ is defined as the angle between the incident plane and the [001] direction on the sample surface. The zero degree refers to the configuration when the incident plane contained the [001] direction of the sample. The incident angle is 45°. The SH intensity is plotted in the radial direction. The intensity is shown in an arbitrary unit shown to the right of each pattern. The four different input and the output polarization combinations are indicated at the top: (a) p-in/p-out, (b) p-in/s-out, (c) s-in/p-out, and (d) s-in/s-out.

Polarization configurations	Experiment	Theory	$\chi^{(2)}_{311}$	$\chi^{(2)}_{222}$	$\chi^{(2)}_{323}$
(a) p-in/p-out					
(b) p-in/s-out					
(c) s-in/p-out					
(d) s-in/s-out					

Figure 3   The decomposition of the SH intensity patterns from 19 nm width Cu nanowires at the fundamental photon energy of 1.17 eV for (a) p-in/p-out, (b) p-in/s-out, (c) s-in/p-out, and (d) s-in/s-out polarization configurations.

configuration (not shown). Next, we assumed a quasi-Cs symmetric Cu nanowires/NaCl(110) system. There are ten independent elements under this system: $\chi^2_{113}, \chi^2_{223}, \chi^2_{311}, \chi^2_{322}, \chi^2_{333}, \chi^2_{211}, \chi^2_{222}, \chi^2_{233}, \chi^2_{121}, \chi^2_{323}$. The calculated patterns reproduce the experiment well as shown in the solid curves in Fig. 1. In Fig. 3 the theoretical decompositions of the patterns of the SH intensity from Cu nanowires of 19 nm width are shown. The contributions from the $\chi^2_{323}, \chi^2_{311}, \chi^2_{222}$ elements are proved to be prominent in p-in/p-out, s-in/p-out, and s-in/s-out polarization configurations, respectively. The SH polarization observed in Fig. 1 could be judged to be in directions 2 and 3 because the first suffices of the three χ^2_{ijk} elements are either 2 or 3. One element χ^2_{222} results from the broken symmetry in the 2: $[1\bar{1}0]$ direction due to the cross-sectional shapes of the nanowires produced by the periodic macrosteps parallel to the 1: $[001]$ direction. It was suppressed by the depolarization effect in Au nanowires produced by the same fabrication technique [1]. The other two elements $\chi^2_{311}, \chi^2_{323}$ in Cu nanowires are found to be different from the prominent $\chi^2_{311}, \chi^2_{323}, \chi^2_{333}$ elements in Au nanowires. We suggest that the suppression of the $\chi^2_{113}, \chi^2_{223}$ elements due to a weak depolarization field in Cu nanowires still lead to the intensity of SHG as low as the noise level in p-in/s-out polarization configuration. The two-lobed patterns in Au nanowires are wider in the middle in p-in/p-out and s-in/p-out polarization configurations due to the contribution from the χ^2_{333} element.

There are two candidate origins of the $\chi^2_{311}, \chi^2_{323}$ elements in Cu nanowires. First, there may be changes of the electromagnetic field gradient normal to the substrate plane due to the plasmon excitation. Second, the electrostatic lightning-rod effect via the geometric singularity of sharply pointed structures may result in electric field enhancement. We are numerically investigating the local electromagnetic field by finite-difference time-domain method in order to clarify the contribution of lightning-rod effect and surface plasmon resonances to the enhancement in our samples. The detailed results will be presented in our next paper [6].

Give your own suggestions

Results and Discussion: Example 4

Results

Figures 1 and 2 show the SH intensity spectra from various wire widths of Cu nanowires and 10 nm Cu nanodots as a function of the SH photon energy, respectively. The SH photon energies are chosen from 2.4 to 4.6 eV. The incident wave vector of the excitation is parallel to the [110] direction of the NaCl (110) faceted substrates. The polarization combinations are (a) p-in/p-out and (b) s-in/p-out, respectively. For the p-in/s-out and s-in/s-out polarization configurations, the signal was very weak and the SH intensity spectra are not shown. The experiment errors are in the range of 10%.

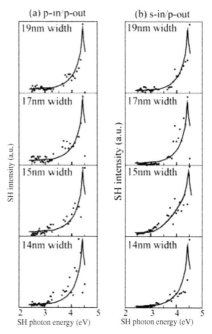

Figure 1 Measured SH intensity spectra from Cu nanowires of different wire widths as a function of the SH photon energy for (a) p-in/p-out and (b) s-in/p-out polarization configurations. The incident wave vector of excitation is parallel to the [110] direction of the NaCl(110) faceted substrates. The solid lines are guide to the eyes.

For p-in/p-out polarization configuration in Fig. 1(a), the SH intensity spectra for Cu nanowires of different wire widths show steady increase above two-photon energy of 3.0 eV. The SH response exhibits a peaked resonance near 4.4 eV. The peaks are similarly observed for s-in/p-out polarization configuration in Fig. 1(b). The spectra show no significant difference for the different wire widths of Cu nanowires.

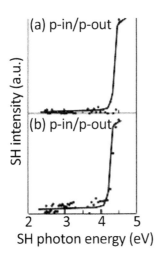

Figure 2 Measured SH intensity spectra from 10 nm Cu nanodots as a function of the SH photon energy for (a) p-in/p-out and (b) s-in/p-out polarization configurations. The incident wave vector of excitation is parallel to the [110] direction of the NaCl (110) faceted templates. The solid lines are guide to the eyes.

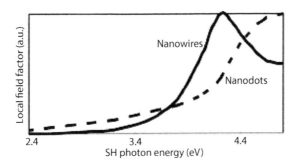

Figure 3 Calculated SHG local field enhancement factor as a function of the SH photon energy for Cu nanowires and nanodots based on Eqs. (1) and (3).

On the other hand, the SH intensity spectra taken from Cu nanodots in Fig. 2 show a nearly flat dependence of SH intensity on the photon energy below two-photon energy of 4.0 eV. The intensity abruptly increases above two-photon energy of ∼4.2 eV, and reaches a maximum at 4.5 eV.

Discussion

The main peak near two-photon energy of 4.4 eV for Cu nanowires in Fig. 1 appears to have a similar origin as the main peak at 4.5 eV for Cu nanodots in Fig. 2. This may be a resonance enhancement of the SH signal with the dipolar mode of the surface plasmon in Cu nanowires. These experimental findings are consistent with the suggestion by Schider et al. that the plasmon excitation in the metallic nanostructures gave rise to the local electric field enhancement [3].

Compare the findings with other study

Next, we perform a theoretical analysis of the SH intensity in order to determine the values of the local field factor and then to find whether the surface plasmon resonance served as the origin of the enhancement of the SH response from Cu nanowires. We calculate the local field enhancement factor for Cu nanowires and Cu nanodots by a quasi-static theory [2]. Since the depolarization factor is zero for the electric field polarized parallel to the major axes, the field enhancement is unity for any frequencies. Therefore, the enhancement is considerably not reduced by the depolarization field.

For the electric field polarized perpendicular to the major axes, the enhancement $L_\perp$ is modified as [2]

$$L_\perp = \left[\frac{\varepsilon_0}{q_\perp \varepsilon_m + (1 - q_\perp)\varepsilon_0} \right]^2. \tag{1}$$

The ε_m and ε_0 are, respectively, the dielectric functions of the metal (Cu) and the surrounding host medium (SiO). The frequency-dependent dielectric constants of Cu are obtained from literature [6]. The depolarization factor $q_\perp$ for the electric field incident along the minor axes b or perpendicular to the major axes a is given by

$$q_\perp = \left[2 + \frac{b}{a} \right]^{-1}. \tag{2}$$

We substitute $b/a = 0.05$ for Cu nanowires and 1.00 for Cu nanodots in Eq. (2), and plot the calculated local field factor, $\left| L_\perp^2(\omega)L_\perp(2\omega) \right|^2$, with respect to the SH photon energy. It is found that the calculated spectra is consistent with the observed SH intensity spectra for Cu nanowires, but it did not reproduce the one for Cu nanodots (not shown). In order to improve the agreement of the calculation with the experimental data, the depolarization factor in Eq. (2) is modified as

$$q_{\perp} = \left[2 + \frac{b}{a}\right]^{-1} - \frac{1}{3}k^2b^2 - i\frac{2}{9}k^3b^3,$$ (3)

with k is equal to $2\pi / \lambda$. The first term on the right-hand side of Eq. (3) is the usual factor due to polarization in the quasi-static limit. The second term comes from the depolarization field that depends on the minor axes b. The third term describes damping effect due to the spontaneous emission of radiation of the induced dipole, and also depends on the minor axes b.

The calculated spectra for Cu nanowires in Fig. 3 still reproduce the experimental findings in Fig. 1. The overall agreement of the spectra of interest between the theory in Fig. 3 and the experiment in Fig. 2 for Cu nanodots improved when we took into account the damping term. It has made the spectra wider and moved the dipolar mode to the higher photon energy. Hence, these theoretical results confirm that the resonant enhancement near the SH photon energy of 4.4 eV originates from the coupling of plasmon resonance in Cu nanowires.

As seen in Fig. 3 the theoretical line-shape of spectra of Cu nanodots is different from that of Cu nanowires because the particle sizes of nanodots are smaller than those of nanowires. When the sizes of particles are small through the reduction of the minor axes b in Eq. (3), the electrons can be accelerated in the presence of the incident light. They, therefore, radiate energy in all directions. Because of this secondary radiation, the electrons lose energy experiencing a damping effect. These results are in agreement with the suggestion of Zeman et al. that the increased electromagnetic damping as the particle sizes decrease leads to smaller enhancement [7].

Compare the findings with other study

Results and Discussion: Example 5

First we have tried to confirm the key part of the total light algorithm. In this measurement, we used the ICG-Intralipid and Intralipid solutions for the target. We define a zero-lifetime fluorescence flux $\Phi_{fluo}^*(t)$ by $\Phi_{fluo} = \int dt' \Phi_{fluo}^*(t') F(t-t')$ where $\Phi_{fluo}(t)$ and $F(t)$ are the fluorescence temporal profile and the fluorescence decay function, respectively. Then, the intensity of the excitation light without the fluorophores can be expressed by $\Phi_{exc}^0(t) = \Phi_{exc}(t) + 1/\gamma \Phi_{fluo}^*(t)$ where $\Phi_{exc}(t)$ is the excitation light intensity with the fluorophores. $\Phi_{exc}^0(t)$ is called total-light in Marjono's paper [3].

Figure 1(a) shows the experimental results with and without ICG in the target. The excitation light was injected at position 0° and data were taken at positions 90° and 180°. The temporal profiles at 180° are always delayed and broader than those at 90° because the geometrical distance becomes larger than that at 180°. The excitation light with ICG decays much faster than that of without ICG due to the absorption of ICG. This difference is corresponding to the energy absorbed by ICG along the excitation light path and this energy is converted to fluorescence. The fluorescence profiles are much broader than the excitation ones due to the lifetime of the ICG-Intralipid complex.

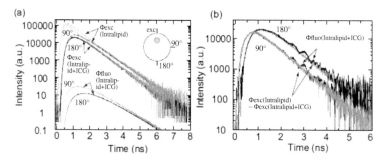

Figure 1 Typical temporal profiles with/without ICG (a) and comparison of the zero-lifetime profile of emission and the difference between profiles of the excitation light with/without ICG (b). The black and gray lines of the excitation profile are measured with ICG and without ICG, respectively in (a). The black and gray lines in (b) are the zero-lifetime and the difference of the profiles, respectively.

The zero-lifetime profile was obtained by the deconvolution with an assumption of a 500 ps single exponential fluorescence decay. The difference of the profiles was calculated by simple subtraction of the excitation profiles without and with ICG. The contribution of the instrumental response function (100 ps) in excitation and fluorescence profiles can be assumed same in our experimental condition and thus the data was not deconvolved by the IRF. Then, the vertical scale was arbitrarily chosen to compare both profiles. The difference and the zero-lifetime profile are in good agreement, as shown in Fig. 1(b). Therefore, the relationship $\Phi_{exc}^{0}(t) - \Phi_{exc}(t) = \alpha \Phi_{fluo}^{*}(t)$ can be confirmed in our system. We can correct the scale factor α, which is determined by detection efficiency of the excitation and emission and the quantum efficiency of the fluorophore. These values will be corrected in the reconstruction.

We will show a reconstructed image from the temporal profile data by total-light algorithm. The procedure of reconstruction was following. First, the raw data were deconvolved by IRF. Then, the fluorescence data was also deconvolved by a single exponential decay with 500 ps lifetime. To make total-light, the emission intensity was calibrated by the filter attenuation and then add the excitation temporal profile with the assumption of the quantum yield (0.03). The sample was excited from the positions 0°, 90°, and 180° and the temporal profile was measured at 30° each. The symmetrical points were not measured but copied in the reconstruction (for example, in case of 0° excitation, data at the points 30°, 60°, 90°, 120°, 150°, and 180° were actually measured. In the reconstruction, these data were duplicated for the other points). We have employed a conjugate gradient method to minimize the deviation of the mean transit time between the measurements and the forward FEM calculation to reconstruct the image, which will be discussed elsewhere.

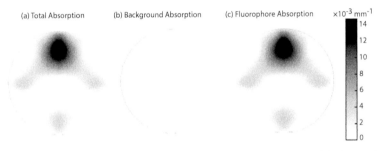

Figure 2 The reconstructed images of 1 μM ICG-Intralipid target. (a) The total absorption obtained by the excitation temporal profiles. (b) The background absorption obtained by total-light. (c) The absorption of the fluorophore (ICG) target obtained by the subtraction of (a) and (b).

Figure 2 shows the result of reconstruction. The total absorption image (a) is just normal DOT image. The peak absorption is about 0.014 mm^{-1} and about 40% of given absorption (0.035 mm^{-1}). This is probably the result of the reconstructed image size; the reconstructed image widely distributes and the peak absorption becomes low. This problem might be improved by using higher moments of the temporal profiles. The background image (b) is almost homogeneous and does not show any shape corresponding to the target. This is consistent with the absorption difference between Intralipid solution and POM. The value of absorption distributes from 0 to 0.6×10^{-3} mm^{-1} and almost agrees with that of POM (0.2×10^{-3} mm^{-1} at 780 nm). Therefore, the total-light calculation can eliminate the target absorption and successfully visualize only the background absorption. The final image (c), which is only determined by the fluorophore absorption, is clearly showing the target but the accuracy is almost determined by the original image (a) reconstructed by the excitation light.

Give your own suggestions

Results and Discussion: Example 6

Results

The fluorescence decay functions with and without the target are shown in Fig. 1. The inset shows the definition of the measurement points. The decay profiles were calibrated by the efficiency of the fiber collections, attenuation of the neutral density filters, accumulation time, and the offset of time origin. The time origin was defined by the time at the peak position of the instrumental function at the injection point. The decay curves almost exponentially decay beyond 1 ns with the abdominal side excitation at C. Since the temporal spreading of the excitation light was not so large (about 100 ps), the decay is almost determined by the fluorescence decay of ICG in Intralipid solution. On the other hand, the peak position of the curve is slightly shifting with the different detection point. The shift is almost determined by the difference of the distance from the excitation point. The decay curves without the target were noisy but still measurable near the excitation point. The peaks of these curves were slightly earlier than those with the fluorescence target. This is most probably due to the contamination from the scattered excitation light by the holder. The fluorescence intensity with target is sufficiently higher than that without the target with the excitation at C. Therefore, the curves are not tainted by the contamination of the background.

In contrast, with the back side excitation at H, the decay curves became more noisy because the fluorescence intensity was significantly smaller than that of the abdominal side excitation at C. At the point G, although the curve is better than others, the shape seems to be deformed by the contamination. Actually, the decay curve without the target also shows a similar shape and this suggests to a large contamination. At the point B, the curve is noisy but almost exponentially decays like the abdominal side excitation. The curve corresponding to B without the target is also plotted in the figure but not visual because the signal was too weak. Therefore, the curve with the target at B is not tainted by the background.

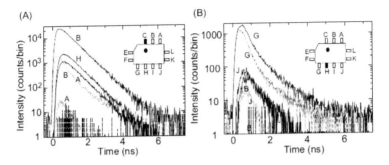

Figure 1 Fluorescence decay functions of some of the pairs of the measurement with and without the target. (A) The profiles with the excitation at a point C near to the target. (B) The profiles with excitation at a point H near the middle of back. The thicker and thinner lines show the curves with and without the target, respectively.

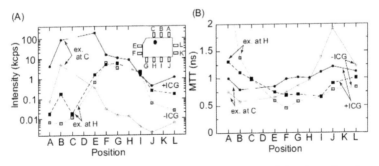

Figure 2 (A) Fluorescence intensity and (B) the MTT of the decay functions with respect to the detection point with the excitation at C (circles) and at H (boxes), with (filled) and without (open) the target.

The fluorescence intensity and the MTT of the decay function with and without the target are shown in Fig. 2. The fluorescence intensity is sufficiently high with the abdominal side excitation at C. In contrast, the fluorescence intensity is significantly low with the excitation at H and the intensity difference between with and without the target becomes smaller. The difference between the intensity with and without the target is largest near the target position at B and C. The MTTs are significantly large at the point with very low intensity less than 10 counts/bin/sec because of the insufficient statistics of data. The MTT with the target is longer than that of the background. The MTT is significantly longer than the fluorescence lifetime of ICG in Intralipid solution because of the propagation time to the detection point and the temporal spreading due to the multiple scattering. The reproducibility of the fluorescence intensity was not good. This is because of the difference in the fiber-skin contact and the positioning of the fiber, which significantly affect the intensity. Therefore, we chose only the MTT as a parameter of the reconstruction.

For the reconstruction, the MTT of total-light should be calculated. The MTT of total-light is given by a simple arithmetic of the means of the excitation temporal response and zero-lifetime functions as $\langle t \rangle_{\text{total}} = (\langle t \rangle_{\text{exc}} \langle I \rangle_{\text{exc}} + \alpha/\gamma \langle t \rangle_{\text{zero-lifr}} I_{\text{fluo}})/(I_{\text{exc}} + \alpha/\gamma I_{\text{fluo}})$, where I_x, α, γ are the intensity of the excitation or emission, the efficiency of the detection, and the quantum efficiency of the fluorophore, respectively. In the calculation of means, the means of the convolution integral become a simple sum of mean of each function in the integral. Thus, the mean of the zero-lifetime function is given by $\langle t \rangle_{\text{zero-life}} = \langle t \rangle_{\text{obs}}^{\text{em}} - \langle t \rangle_{\text{instrum}} - \langle \tau_{\text{fluo}} \rangle$, where $\langle \tau_{\text{fluo}} \rangle$ is the average lifetime of the fluorophore and the suffixes "obs" and "instrum" indicate the mean of the observed function and that of the instrumental functions, respectively. Similarly, the mean of the excitation temporal response function is calculated by $\langle t \rangle_{\text{exce}} = \langle t \rangle_{\text{obs}}^{\text{exc}} - \langle t \rangle_{\text{instrum}}$. Here, we assume the parameters as follows; $\gamma = 0.1$, $\langle \tau_{\text{fluo}} \rangle = 0.5$ ns, and $\alpha = 6.8$.

Figure 3 shows the MTT of the excitation light and the ratio of the MTT of total-light and excitation light with respect to the detection point. The ratio is a parameter of the energy loss by the fluorescence target with a specific pair of excitation and detection points. The deviation in the ratio from unity is the origin of the visualization of the absorption of fluorophores by the total-light algorithm. The deviation was about 2% of the MTT of the excitation light. With the excitation at C, the contribution in the MTT of the excitation light from the absorption of the fluorophores can be observed throughout all detection points. On the other hand, this can be found at the only few points close to the target with the back side excitation at H. This is because that the observed emission intensity at the back side detection and excitation was high but negligibly small in comparison to the excitation intensity. Therefore, the strongly deformed curve at G in Fig. 1 will not affect the reconstruction.

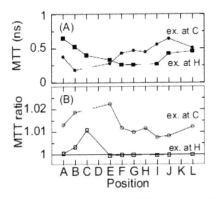

Figure 3 (A) MTT of the excitation light and (B) the ratio of the MTT of total-light and excitation light with respect to the detection points with the excitation at C (circles) and H (boxes).

The reconstructed images are shown in Fig. 4. The left side of the reconstructed image was deformed from the rectangular shape to account a gap between the holder and the animal. The location of the target is schematically drawn in the figure. The scattering coefficient was assumed to be 1.0 mm^{-1} in the reconstruction, and the initial value of the absorption was homogeneously distributed in the cross section with 0.01 mm^{-1}. The number of the nodes of the FEM mesh was 961 points. The middle of the cross section is visualized as a relatively higher absorption region, but there is no particular distribution of the higher absorption region except the center of the image. This high absorption spot around the center is due to the artifact of the image reconstruction. The image reconstructed by total-light is very similar to the image reconstructed by the excitation light. This is because of the small difference in the MTTs. However, the difference between (A) and (B), which is corresponding to the image of the absorption by the florophores, clearly shows a high absorption region in which the target is expected to exist as shown in Fig. 4(C). The image quality is still not good, but the difference between the conventional DOT and the total-light images can visualize the target, which cannot be distinguished in the conventional DOT image.

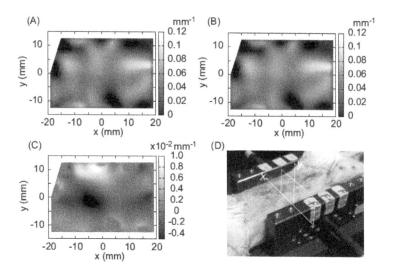

Figure 4 The absorption images of rat in vivo. (A) The absorption image of the cross section with the target reconstructed by the excitation light. (B) The absorption image without the target reconstructed by total-light. (C) The absorption image of the fluorophores obtained by the subtraction between images (A) and (B). (D) The definition of the coordinate and location of the target.

Discussion

The most difficult part of this approach is to estimate the intensity ratio of the excitation and emission light. The intensity is totally dependent on the experimental conditions, such as the optical coupling between the collection optics and the tissue surface and the contamination. Fortunately, the ratio is a relatively robust parameter if the detection of the excitation and emission are conducted with the same optics. Therefore, if the efficiency ratio of the detection at the excitation and emission could be estimated carefully, it may not have the large problem. On the other hand, the change in the spectrum of the fluorophores under the in vivo conditions may cause the problem in the estimation of the ratio. The correction efficiency of the emission is varied with the spectrum shift because the emission filter selects a part of the emission spectrum. In addition, the spectrum change is a signature of change in the optical properties of fluorophore, such as the quantum efficiency and the lifetime.

These changes will violate the assumption in the total-light approach. Here, we have carefully estimated with the ICG-Intralipid mixture, which is more stable than ICG aqueous solution. But, there still may be some errors. The stable fluorophore dye is required for much clear information in the total-light approach. We recently have found a very stable new complex of IR806, which is a cyanine dye similar to ICG, and serum albumin.[7] This kind of dye or dye complex may be applicable in future applications.

The next problem is failure of the assumption. The wavelength dependence of scattering will not strongly affect. However, the absorption is more serious problem. The difference of the absorption coefficients at the excitation and emission wavelengths can be eliminated when a closer wavelength pair can be used. However, in practice, the closer wavelength pair will cause the contamination from the excitation light and the minimization of the contamination distortion gives higher priority in the selection of the wavelength pair. A higher concentration of the dye also causes the re-absorption effect of the emission light; the emission light will be more attenuated than the expectation without the re-absorption effect, resulting the intensity and the MTT of total-light will be underestimated. This will be solved by the several trials with the modification of the MTT or the efficiency parameter α.

The total-light approach is a unique method. This algorithm may technically belong to a variety of the conventional DOTs using fluorophore. The target image is finally generated from the subtraction of the images with and without targets. This subtraction process eliminates the complexity due to the unknown inhomogeneous absorption of the tissue and clearly visualizes the fluorophore's absorption. This property is expected to be extremely useful in vivo image reconstruction.

Give your suggestions

Results and Discussion: Example 7

Figure 1(a) shows the intensity of the reflected beam from the reference arm of the low-coherence reflectometer during scanning. The uniformity of the intensity during scanning verifies that there was no walk-off effect in our system. The indentations occurred at the tiled angles around 0 and 10 degrees represent the effects beyond the allowable scanning range. Figure 1(b) shows the interference signal when a He–Ne laser was used as the light source.

Show improvement

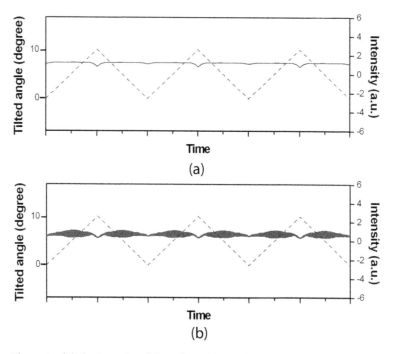

Figure 1 (a) The intensity of the reflected beam from the reference arm of the low-coherence reflectometer during scanning. (b) The interference signal when a He–Ne laser was used as the light source.

To demonstrate the performance of the low-coherence reflectometer, various samples were introduced for scanning. Figure 2(a) shows the interference signal of a reflection mirror. The dynamic range of the low-coherence reflectometer was estimated to be 20 dB. It was mentioned that the scanning depth of the optical delay line could achieve 2.9 mm when the vibration angle of the scanning mirror was 9.6 degrees. When the distance between the retro-reflector and the scanning mirror is reduced, the allowable scanning angle is increased and the scanning range can be further improved. Figure 2(b) shows the interference signal when two slides were introduced as the sample. The thickness of one slide was measured to be about 1.1 mm, thus the scanning range larger than 3 mm was achieved when the refractive index of the slide was assumed to be 1.5. Figure 3 shows the two-dimensional image of a stack of 11 coverslips.

Show application

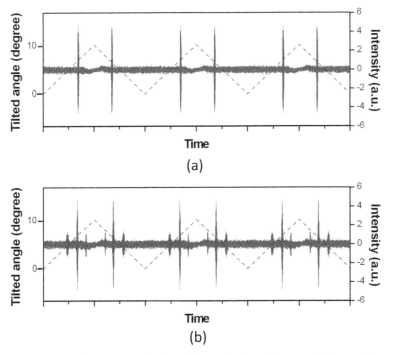

Figure 2 Interference signals when (a) a reflection mirror and (b) two slides were used as samples.

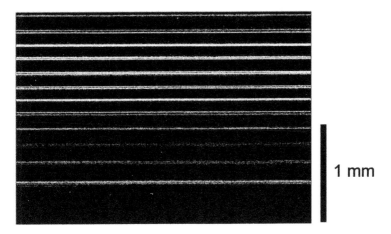

Figure 3 Two-dimensional image of a stack of 11 coverslips. The refractive index was assumed to be 1.5.

Results and Discussion: Example 8

In Fig. 1(a) we show the intensity of the reflected beam from the reference arm of the low-coherence reflectometer during scanning. The uniformity of the intensity during scanning verifies that there was no walk-off effect in the system. The indentations occurred at the tiled angles around 0 and 10 degrees represent the effects beyond the allowable scanning range. On the other hand, Fig. 1(b) shows the interference signal when a He–Ne laser was used as the light source.

> Show improvement

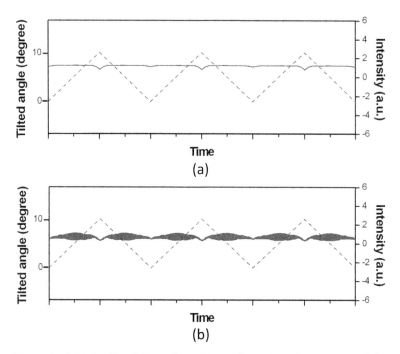

Figure 1 (a) Intensity of the reflected beam from the reference arm of the low-coherence reflectometer during scanning (line) and the tilted angle of the scanning mirror (dashed line). (b) Interference signal when a He–Ne laser was used as light source (line) and the tilted angle of the scanning mirror (dashed line).

In order to show the performance of the low-coherence reflectometer, various samples were introduced for scanning. Fig. 2(a) shows the interference signal of a reflection mirror. The dynamic range of the low-coherence reflectometer was estimated to be 20 dB. It was mentioned that the scanning depth of the optical delay line could achieve 2.9 mm when the vibration angle of the scanning mirror was 9.6 degrees. When the distance between the retro-reflector and the scanning mirror is reduced, the allowable scanning angle is increased and the scanning range can be further improved. On the other hand, Fig. 2(b) shows the interference signal when two slides were introduced as the sample. The thickness of one slide was measured to be about 1.1 mm, thus the scanning range larger than 3 mm was achieved when the refractive index of the slide was assumed to be 1.5. Fig. 3 demonstrates the two-dimensional image of a stack of 11 coverslips.

Show application

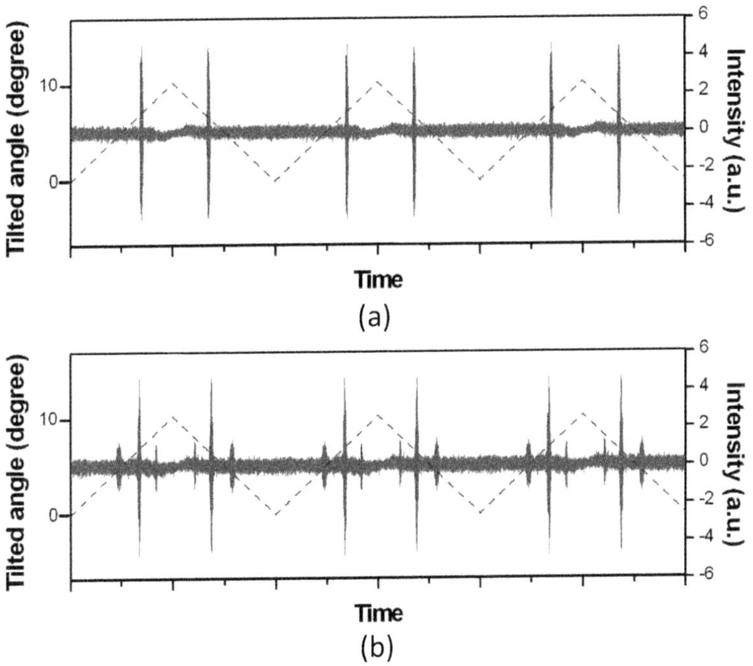

Time

(a)

Time

(b)

Figure 2 (a) Interference signals when a reflection mirror was used as samples (line) and the tilted angle of the scanning mirror (dashed line). (b) Interference signals when two slides were used samples (line) and the tilted angle of the scanning mirror (dashed line).

Figure 3 2-D image of a stack of 11 coverslips. The refractive index is 1.5.

Results and Discussion: Example 9

Structural properties. Figure 1 showed the XRD diffraction patterns of the quartz substrates and the AlN films grown under nitrogen concentrations (CN) of 20%, 40%, 60%, 80%, and 100%. For the AlN films deposited at 20% CN, three peaks were observed. The first peak occurred at 36.5° with a full width at half-maximum (FWHM) of 0.29° corresponding to the diffraction from the (002) plane in the film. The second peak centered at 50.3° with a FWHM of 0.29° corresponding to the diffraction from the (102) plane in the film. The last peak existed at 66.38° with an FWHM of 0.45° corresponding to the diffraction from the (103) plane in the film. The (002) and (103) planes were indexed on the basis of the hexagonal wurtzite-type structure. These indicated the preferential AlN growth orientation along the c-axis perpendicular to the surface substrate.

In general, the (103) plane occurred for all % CN and its intensity decreased when increasing CN from 20% to 40%. After that its intensity increased when increasing CN from 60% to 100%. On the other hand, when increasing CN from 20% to 40%, the intensity of the (002) peak increased to reach a maximum. After that the (002) peak intensity decreased when increasing CN from 60% to 100%. As the intensity of the (002) diffraction peak was the most intense at 40% CN, this was judged to be an improvement of our AlN film's crystallinity.

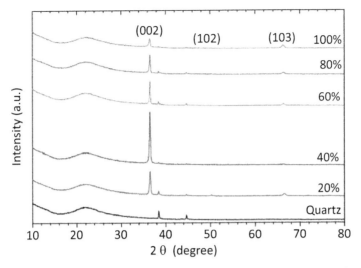

Figure 1 XRD diffraction patterns of the quartz substrates and the AlN films grown under various nitrogen concentrations.

Moreover, the formation of the (002) plane in our AlN film was in agreement well with the formation energy from Kao et al. [9] showing formation of this plane has higher energy than that of others. That was, when increasing CN up to 40%, the probability of collision between Al-sputtered atom and nitrogen atom increased while the probability of collision between Al-sputtered atom and argon atom decreased. This resulted in reduction of the energy loss of the sputtered particles. The sputtered aluminum atom arriving the substrates with high energy induced the formation of the (002) plane in the AlN film. After that, when CN was higher than 40%, the nitrogen atoms were served as the sputtering gas. The lighter sputtering atoms caused the lowering of Al-sputtered yield resulting in a gradual reduction in the formation of (002) plane.

The morphologies of the deposited films under different CN were shown in Fig. 2. The root mean square (RMS) roughness values of the deposited film under CN of 20%, 40%, 60%, 80%, and 100% were 2.20, 2.72, 2.06, 2.64, and 2.90 nm, respectively. It could be inferred from the AFM data that the surfaces of all films were of comparable smoothness and the dense granular structure was observed except for the CN of 20% (AFM image of the AlN film deposited under CN of 80% was not shown here). The grain sizes tended to decrease significantly when increasing CN.

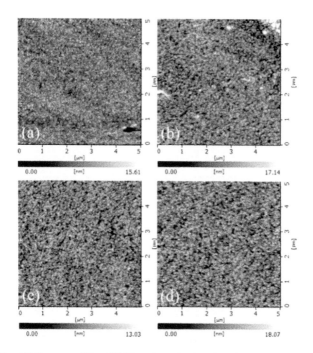

Figure 2   AFM images of the AlN films deposited under nitrogen concentrations of (a) 20%, (b) 40%, (c) 60%, and (d) 100%.

Optical properties. Figure 3 showed the optical transmittance spectrum of the AlN thin films deposited on the quartz substrates under CN of 20%, 40%, 60%, 80%, and 100% as a function of wavelength. Interference pattern in the transmission spectrum showed the smooth reflecting surfaces and not much scattering loss at these surfaces. Also, the films showed good transparency in the visible region (Transmission T is about 0.80).

Table 1   The fitting parameters of two-term Cauchy dispersion relation, the thickness calculated by envelope method, and the thickness measured by FE-SEM of the deposited films under CN of 20%, 40%, 60%, 80%, and 100%

C_N [%]	a	b	$d_{envelope}$ [nm]	d_{FE-SEM} [nm]
20	1.988	1.968	784 ± 6	774
40	1.982	1.764	1153 ± 9	1161
60	1.991	1.791	1204 ± 8	1240
80	1.970	1.815	1063 ± 9	1031
100	1.961	1.892	824 ± 10	863

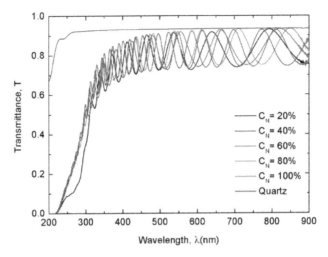

Figure 3 Optical transmission spectrum of the AlN thin films deposited on the quartz substrates under nitrogen concentrations of 20%, 40%, 60%, 80%, and 100%.

All films showed a sharp decrease in transmittance at wavelength range from 200 nm to 300 nm partly due to absorption. In the medium-to-weak absorption region according to Swanepoel's method [10,11], the refractive index of the film (n), absorption coefficient (α), and the film thickness (d) could be estimated. The refractive index was fitted by using a two-term Cauchy dispersion relation: $n = a + b/\lambda^2$. The values of a and b of the films were listed in Table 1. It was noted that the correlation coefficients of the fitted dispersion relation of all films were more than 0.992. The FE-SEM was used to cross-check the thickness determined from the envelope method. We could see that the values of thicknesses determined by the envelope method were close to that of thicknesses measured by FE-SEM.

At the wavelength of 550 nm, the refractive index of the film was 2.05, 2.04, 2.05, 2.03, and 2.02 under CN of 20%, 40%, 60%, 80%, and 100%, respectively. This indicated that the refractive index weakly depended on CN. However, our calculated refractive indices were in the same range as reported in the AlN films deposited by the reactive magnetron sputtering in Ref. [12].

By using Tauc's relation: $\alpha E = A(E - E_g)^{0.5}$, where A is a constant, E is the photon energy, and E_g is the films

> Compare the findings with other study

optical band gap, the Tauc's plot; plot of $(\alpha E)^2$ vs. photon energy (E) was represent in Fig. 4(a). The optical band gap of the films orientation along the c-axis perpendicular to the surface substrate could be determined at photon energy between 5 eV and 6 eV. The optical band energy of the film was 5.60, 5.67, 5.63, 5.62, and 5.57 eV under CN of 20%, 40%, 60%, 80%, and 100%, respectively. This indicated that optical band gap energy strongly depended on CN. In Fig. 4(b), when increasing CN from 20% to 40%, the energy increased. In contrast, when increasing CN from 60% to 100%, the energy decreased. Our calculated energies were in the same range as reported in the AlN films deposited by the reactive magnetron sputtering in Ref. [12].

Compare the findings with other study.

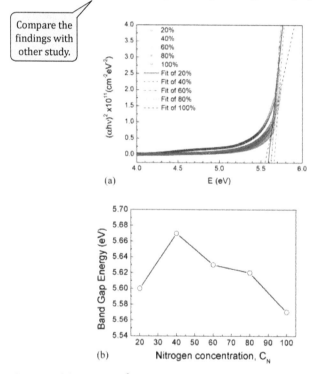

(a)

(b)

Figure 4 (a) Plot of $(\alpha E)^2$ vs. photon energy (E) and Tauc's plot of each condition for the AlN films deposited under various %CN. (b) The optical band gap of the AlN films as a function of CN.

The intensity of the Al peak at 395.6 nm also could be used to identify our film properties when the aluminum sputtered atom was monitored by optical emission spectroscopy (OES). The aluminum peak (at wavelength of 395.6 nm) on optical emission intensity of the film under CN of 20%, 40%, 60%, 80%, and 100% CN was 225, 321, 285, 270, and 180 a.u., respectively. This meant that when increasing CN from 20% to 40%, the peak intensity increased. In the contrary, when increasing CN from 60% to 100%, the intensity decreased. This behavior was in agreement with our data taken from the band gap energy and the XRD (002) peak intensity.

Results and Discussion: Example 10

Effect of Nitrogen Concentration. The effect of nitrogen concentration on the growth rate of the AlN films was shown in Fig. 1. The growth rate of the films was calculated from the deposition time and film thickness. The obtained films thickness was varied between 774 nm and 1240 nm. It was shown that increasing CN significantly reduced the film growth rate due to the lowering of sputtering yield for nitrogen ions as compared with argon ions [6]. In addition, the increasing CN causes the target surface to more cover with the reaction products (AlN). Consequently, the sputtering yield of the target decreased with increasing the reaction products on the target surface because aluminum nitride has higher surface binding energy than aluminum [7]. This behavior was in agreement well with many literatures [6,8,9].

Compare the findings with other studies.

Bonding Formation. The bonding formation of the AlN films was studied by FTIR and Raman spectroscopy. The effect of CN on the peak position was not observed. In Fig. 2(a), FTIR peak of the films was confirmed near 667 cm^{-1} corresponding to E1 (TO) vibration mode of the infrared active Al–N bond [10]. The peaks were likely to shift from an original absorption band of 670 cm^{-1}, partly due to the residual stress in the films caused by the sputtering process [11]. Fig. 2(b) showed Raman spectra of the films. The Raman peak near 607 cm^{-1} and 651 cm^{-1} represented A1 (TO) and E2 (high) vibration mode of the AlN materials [10], respectively. The E2 (high) peak was likely to shift to the lower frequency due to the residual stress in the films induced by the sputtering process and thermal expansion coefficient between the films and the quartz substrate [10–12]. It was noted that there were also two additional broaden band around 450–500 cm^{-1} and 740–840 cm^{-1} due to the local vibration mode of the quartz substrates [13].

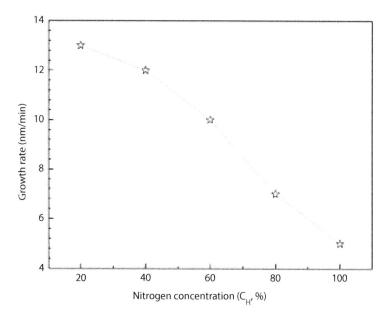

Figure 1   Growth rate of the AlN films as a function of nitrogen concentration.

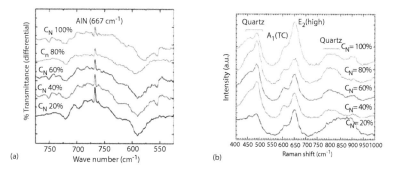

(a)

(b)

Figure 2   FTIR (a) and Raman (b) spectra of the AlN films with different nitrogen concentrations.

Crystal Orientation and Structure. Figure 3(a) showed the XRD patterns of a quartz substrate and the AlN films with different nitrogen concentrations. For the film with CN of 20%, three peaks were observed. The first peak occurred at 36.5° corresponding to the diffraction from the (002) plane in the film. The second peak centered at 50.3° corresponding to the diffraction from the (102) plane in the film. The last peak existed at 66.38° corresponding to the diffraction from the (103) plane in the film. The (002) and (103) planes were indexed on a basis of the hexagonal wurtzite-type structure. The film with CN of 40% showed the highest intensity of the (002) plane, indicating the best preferential AlN growth orientation along the *c*-axis perpendicular to the surface substrates. Figure 3(b) showed the cross-sectional FE-SEM image of the film with CN of 40%, the columnar structure is clearly visible, confirming the highly oriented columnar crystals perpendicular to the substrates. Such films structure was also confirmed in Ref. [14].

Compare the findings with other study.

Surface Morphologies. Figure 4 showed the AFM images of the AlN films under different nitrogen concentrations. The root-mean-square surface roughness of the films was found to be in a range of 1.86–2.87 nm. The dense columnar grains were uniformly observed on the surface of the films under all CN conditions, except for CN of 20%. The grains size of the films was found to be in a range of 61–85 nm and tended to decrease with CN.

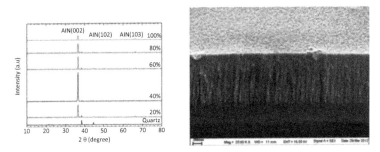

Figure 3 XRD pattern (a) of the AlN films with different nitrogen concentrations and cross-sectional SEM image (b) of the AlN film with CN of 40%.

Bulk Resistivity. For the AlN films to serve as piezoelectric materials in MEMs and SAW devices, the resistivity of the oriented *c*-axis films must be high [14]. The metal–insulator–metal (MIM) structures were formed by using our fabricated AlN insulator under our optimized films condition of 40% CN with a thickness of approximately of 1 μm. The film was sandwiched between two square-aluminum electrodes with area of 1 cm^2. Fig. 5 showed the relation between current density (positive and negative bias) and electric field. It also showed bulk resistivity with respect to electric field. It was seen that the current density was below 100 pA/cm^2 when the electric field was below 100 kV/cm. The calculated bulk resistivity of the film was then in a range of 7×10^{14} to 2×10^{15} Ωcm. This resistivity was greater than those reported in Refs. [14,15], while it was very close from one in Ref. [16]. This resistivity was also close to a single crystal AlN in Ref. [17], acceptable for the piezoelectric materials in SAW devices.

Compare the findings with other studies.

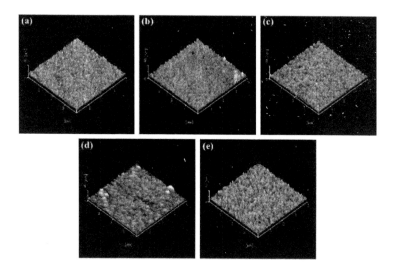

Figure 4 AFM images of the AlN films under nitrogen concentrations of (a) 20%, (b) 40%, (c) 60%, (d) 80%, and (f) 100%.

Piezoelectric Property. Piezoelectric property of the film was studied by Michelson interferometer. The said MIM structure was used to study the inverse piezoelectric effect. A modulation voltage was applied between the electrodes of the piezoelectric film. The mirror-like top aluminum electrode was oscillated at the same frequency as the applied voltage. In the Michelson interferometer method, the bias-induced deformation was detected in term of intensity of interference fringe by photodetector [18]. The vertical displacement (Δd) of the sample surface was calculated by the relation: $\Delta d = V_{out}\lambda/2\pi V_{pp}$. The output voltage ($V_{pp}$) was calculated from $V_{pp} = V_{max} - V_{min}$, where V_{max} and V_{min} are voltages corresponding to maximum and minimum fringe intensities, respectively. The measured voltage corresponding to Δd was defined as V_{out}. After that, the clamped piezoelectric coefficient

($d_{33,f}$) was calculated by the relation: $d_{33,f} = \Delta S_3/E_3 = \Delta d/V$, where the strain change along c-axis ΔS_3 is $\Delta d/t$, electric field along the c-axis E_3 is V/t, and V is applied voltage [19].

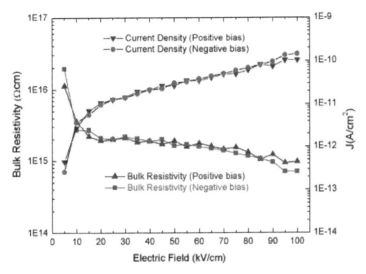

Figure 5 Current density–electric field (J–E) plot and calculated bulk resistivity as function of electric field of the MIM structures with 40% CN.

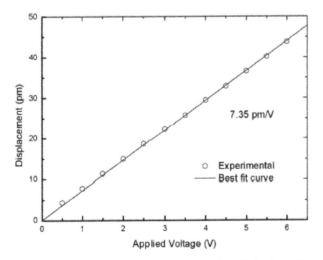

Figure 6 Displacement amplitudes as a function of applied voltage (sinusoidal wave at 1 kHz) for the MIM structures with 40% CN.

Figure 6 showed a linear dependence of the vertical displacement of the electrode surface as a function of the applied voltage. The calculated clamped piezoelectric coefficient ($d_{33,\,f}$) measured by Michelson interferometer was 7.35 pm/V in average. We then estimated the normal unclamped piezoelectric coefficient (d_{33}) by the following relation

$$d_{33} = d_{33,f} - 2\left(\frac{d_{31}S_{13}^{E}}{S_{11}^{E} + S_{12}^{E}} \right) \tag{1}$$

After we used the compliance value ($S_{11}^{E}, S_{12}^{E}, S_{13}^{E}$) and d_{31} in Ref. [19], and the calculated d_{33} was obtained at 11.03 pm/V. This value was found to be higher than ones from Refs. [19–21].

Compare the findings with other studies.

Results and Discussion: Example 11

In Fig. 1 we show the experimental results of the optical path difference as a function of the tilted angle of the scanning mirror in order to demonstrate the performance of our optical delay line. These experimental results are estimated at positions of the maximum interference signals. It is found that within the maximum tilted angle of the scanning mirror of approximately 10°, we obtain the imaging depth represented by the optical path difference at about 3 mm.

Show improvement

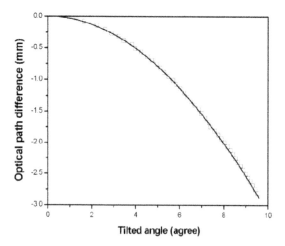

Figure 1 Optical path difference as a function of the tiled angle of the scanning mirror.

Results and Discussion: Example 12

The scanning electron microscope (SEM) image of a typical tungsten nanotip made by a drop-off etching method was shown in Fig. 1(a). The magnified SEM image of the same tip, showing smooth surface of the tip with an average radius of about 24 nm, was shown in Fig. 1(b). We have fabricated over ten tips by using this etching method, and confirmed that the tip shape was controllable by using our home-made etching process controller. The success rate in making the sharp tips having an average radius of about 24 nm at the apex was over 80%.

Show improvement

Next, the field emission characteristic of the fabricated tips was analyzed based on the Fowler–Nordheim (F–N) theory. According to the F–N theory, the field-emission current density J (emission current/anode area) was related to the applied voltage E and the work function ϕ (4.5 eV for tungsten) by the relation [9]

$$J = \frac{1.6 \times 10^{-6} E^2}{\phi} \exp\left(-\frac{6.9 \times 10^9 \phi^{3/2}}{E}\right) (\mathrm{A\,m^{-2}}) \qquad (1)$$

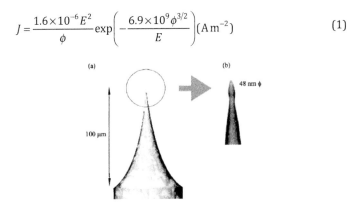

Figure 1 SEM images of the tungsten nanotip with (a) low and (b) high magnification after electrochemical etching in an aqueous NaOH solution.

Figure 2 was field emission data taken with a tungsten tip radius of 24 nm. In Fig. 2(a) we showed the current density J as a function of the applied voltage E, both with and without the laser illumination. In both cases, data were fit well by the F–N equation according to Eq. (1). There was also a clearer exponential increase of the current density with increasing the applied voltage once it exceeded the turn-on field which was estimated to be about 5 V/μm to achieve a current density of 500 A/m². On the other hand, a typical F–N plot was a plot of ln

(J/E^2) versus ($1/E$), which should yield a straight line for a field emission phenomenon. Figure 2(b) showed the corresponding F–N plot of the tip. The straight lines was generally observed, indicating that field emission from the tips was a barrier-tunneling, quantum-mechanical process.

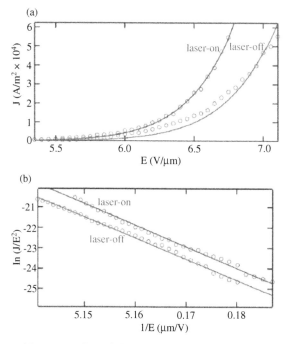

Figure 2   Field emission data of the tungsten nanotip; (a) current density (J) as a function of the applied voltage (E) and (b) corresponding Fowler–Nordheim (F–N) plot. A fit to the data was shown by the solid line.

Because the dimensions of the tips were smaller than the laser wavelength, the field enhancement could take place. The field enhancement factor β correlated the applied voltage E with the local electric field F at the emitting surface by the relation [10]

$$F = \beta E. \tag{2}$$

This field enhancement factor could be calculated from slope of the F–N plot in Fig. 2(b). By analyzing the data shown in Fig. 2(b), the field enhancement factor β was estimated to be 500. According to

Eq. (2), when the applied voltage E was equal to 5 V/μm, the local electric field strength at the tip F was equal to 2 V/nm in order to achieve and start an appreciable tunnel current from our tips.

Furthermore, we also compared field emission data of the tungsten tips with and without laser illuminations as seen in Fig. 2. The tip from the laser-on measurement tended to emit a higher field emission current than those from the laser-off measurement. This was because of laser-induced field-emission behavior. We suggested that this excellent field emission performance also might apply as a new technique for producing microwave radiation, and also might show promise as a new type of source for terahertz radiation.

Give suggestions

Results and Discussion: Example 13

Figures 1(a), (b), and (c) show the fluorescence temporal profiles from a 1.0 μM indocyanine green (ICG) target 6 mm in diameter with different target locations, namely d = 4.5, 7.5, and 15 mm, apart from the phantom surface. The difference of the fluorescence temporal profile at symmetrical point B (30°) and D (−30°) became large when the target was near the phantom boundary or point C (0°). This was because the total path length variation between excitation light and emission light in the tissue phantom was increased. Furthermore, the peak intensity and rising edge of fluorescence temporal profile at 30° were higher and earlier than those of the profile at −30°. This result suggested that the front region of the target was more excited than the backside region. Then, the emitted fluorescence traveled a short path length to the detection point. On the other hand, typically the fluorescence intensity would decay as $I(t) = I_0 \exp(-t/\tau)$, where τ is a fluorescent lifetime. By fitting the observed fluorescence-decay profiles to $I(t) = I_0 \exp(-t/\tau)$, one can obtain τ = 0.5 ns.

Next, we characterized the fluorescence temporal profile in terms of fluorescence peak intensity ratio, as shown in Fig. 2. When the target came close to the phantom boundary, the degree of asymmetry in terms of fluorescence peak intensity ratio was further from unity. This indicated that the symmetrical plane of the target with respect to point A (90°) was broken. Moreover, the degree of this asymmetry was increased when the ICG concentration was increased. Increasing the absorption coefficient of ICG enhanced a gradient of emission intensity at the target region.

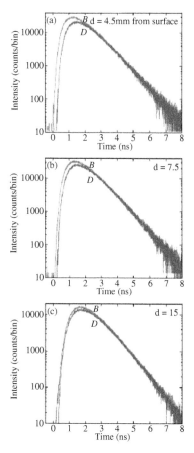

Figure 1 Fluorescence temporal profile as a function of geometrical configuration at different detection points from an indocyanine green (ICG) target of 6 mm in diameter. The target was located at d = 4.5 mm (a), 7.5 mm (b), or 15 mm (c), measured from the phantom surface to the target center. The target concentration was 1 μM ICG solution. The detection points B and D were 30° and −30°, respectively.

These basic results showed a promising approach for fluorescence targeting and further reconstruction of the fluorescence image. The asymmetry of the fluorescence peak intensity ratio was systematically dependent on the target depth and concentration. Therefore, a complete set using this parameter was able to identify an unknown target location. For example, the target size and concentration were known to be 6 mm and 0.5–2.0 mM ICG solution, respectively. If the fluorescence peak intensity ratio was beyond 1.3, an unknown target position was expected to be about 4.5 mm or more from the phantom boundary. By contrast, if the fluorescence peak intensity ratio was below 1.3 (dash line), an unknown target point tended to be deeper than 7.5 mm, apart from the phantom surface.

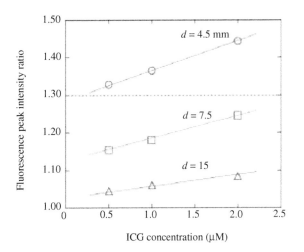

Figure 2 Indocyanine green (ICG) concentration as a function of fluorescence peak intensity ratio at a detection point of 30°/–30°. The target was 6 mm in diameter and was located at d = 4.5 mm, 7.5 mm, or 15 mm, measured from the phantom surface to the target center.

This asymmetrical feature suggests that the fluorescence peak intensity ratio carries geometrical information about the target location. The image quality of the target could then be improved from its fluorescence intensity. On the other hand, our analysis would be useful to decide the excitation–detection geometry and to consider whether this parameter is effective to use in image reconstruction in a tissue model.

Give suggestions

Results and Discussion: Example 14

The transmission electron microscopy images displayed in Fig. 1a and Fig. 1b show uniform gold nanorods and uniform palladium-coated gold nanorods, respectively. The short axis b and the long axis a of the gold nanorod core are 40 ± 3 and 96 ± 6 nm, respectively. The aspect ratio a/b is therefore 2.4 ± 2. The thickness of the coating palladium shell is equal to 3.0 ± 0.2 nm.

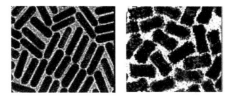

Figure 1 Transmission electron microscopy images of gold (left panel) and palladium-coated gold (right panel) nanorods.

The optical absorption spectra measured for the palladium-coated gold nanorods with the Pd/Au weight ratios 0, 0.05, 0.10, and 0.20 are shown in Fig. 2. Two main plasmon peaks can be detected for all of our samples. The first is linked to electron oscillations perpendicular to the long axis of the rod (a so-called transverse plasmon band), and the second is associated with conduction-electron oscillations parallel to the long axis of the rod (a so-called longitudinal plasmon band). Of those two modes, the transverse oscillations of electrons give rise to the absorption band located at the shorter-wavelength side (at about 525 nm), which corresponds to nanospheres. That location of the resonance is independent of the palladium-shell thickness and the changing Pd/Au weight ratio because the short axis a is dominant for the spherical-like segments. The resonance shows no detectable shift when the Pd/Au weight ratio increases from 0.05 to 0.20. This result can be explained with the approach developed by Sandrock et al. [20] suggesting that the light polarization is weakly affected by the centrosymmetric structures having the aspect ratio $a/b = 1$.

Compare the findings with other study.

On the other hand, the longitudinal oscillations of electrons result in the band located in the longer-wavelength part of the spectrum, because the long axis b is dominant for the case of rod-like segments. The peak position depends on the palladium-shell thickness, thus suggesting that the effect of gold core on the plasmonic sensitivity is completely screened when the palladium shell thickness becomes larger. As a consequence, increasing thickness of the palladium shell imposes a blue shift of the plasmon absorption maximum. This longitudinal plasmon band corresponds to the optical response of the pure palladium nanorods, due to less negative values of the real part of the permittivity of palladium as compared to gold [21]. For instance, the absorption band shifts from 860 to 820 nm when the Pd/Au weight ratio increases from 0.05 to 0.20. This result can also be explained using the results by Sandrock et al. [20], which suppose that the light polarization is strongly affected by the noncentrosymmetric structures having the aspect ratio $a/b \neq 1$. In other words, by varying the Pd/Au ratio, one can adjust the blue shift of the plasmon band to be anywhere beginning from 860 and ending at 820 nm. To study dependence of the plasmon peak on the refractive index of the surrounding medium, we have dispersed the palladium-coated gold nanorods into different organic solvents (The relevant refractive indices are displayed in Table 1). Fig. 3 shows the peak positions of the plasmon resonance for different Pd:Au nanorods immersed into different organic dielectric media. The peak location for the transverse plasmon band is independent of the dielectric medium. This implies that the absorption peak located at 525 nm does not change when the refractive index of the medium increases from 1.33 to 1.49 (not shown in Fig. 3).

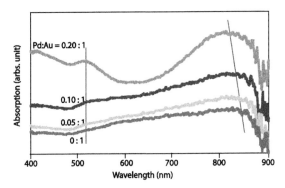

Figure 2 Measured absorption spectra of the palladium-coated gold nanorods with various Pd/Au weight ratios. Left (right) peak represents TPB (LPB). Each sample is immersed in DI water.

By contrast, due to recognizable dependence of the plasmonic sensitivity for the longitudinal mode on the palladium-shell thickness, the palladium-coated gold nanorods with different palladium-shell thicknesses give rise to distinct plasmon-peak wavelengths even in the same surrounding medium. Namely, the plasmon peak wavelengths for the nanorods having the ratios Pd:Au = 0.20:1, 0.10:1, 0.05:1, and 0:1, which are immersed in deionized water (n = 1.33), are respectively equal to 820, 830, 840, and 860 nm. Then, by varying the Pd/Au ratio for the case of the same medium, one can adjust the plasmon-band shift to be anywhere from 820 to 860 nm.

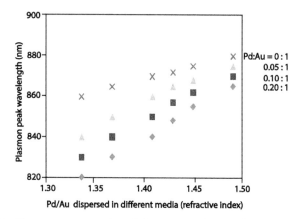

Figure 3 Plasmon peak wavelength (LPB mode) from Pd/Au immersed in the different dielectric media.

Furthermore, the peak position of the longitudinal plasmon band with a given palladium-shell thickness shows evident red shift with increasing refractive index. Namely, the plasmon peak shifts for the cases of Pd:Au = 0.20:1, 0.10:1, 0.05:1, and 0:1 are respectively 820–860, 830–867, 840–874, and 860–880 nm, when each sample is immersed in the medium with different refractive index (1.33–1.49). The shift value depends linearly on the refractive index. The host media with higher refractive indices (i.e., higher dielectric constants) are effectively more polarizable and thus couple more readily with the surface plasmon electrons. The energy needed to collectively excite electrons is then reduced. In this case the maximum of the plasmon absorbance is shifted toward the red side of spectrum.

There are two possible origins of the observed spectral changes associated with the electron oscillations in our samples. First, there is a change in the electromagnetic field gradient normal and/or parallel to the sample plane, due to the plasmon excitation. Second, as seen from Fig. 1, an electrostatic lightning-rod effect via geometric singularity of sharply pointed structures may result in enhancement of the electric field. Numerical investigations of the local electromagnetic fields with the finite-difference time-domain technique are now in progress. They can unravel the nature of the lightning-rod effect and

Give your own suggestions

plasmon modes enhancing in our samples. The more detailed results will be presented in our forthcoming article.

In summary, both the presence of the palladium shells deposited onto the Au-core nanorods and the changes in the refractive index of the surrounding dielectric medium are marked by a change in the sample colour and a shift in the plasmon-peak position proportional to the changes in the electromagnetic field located around the nanorod surfaces. The UV-Vis-NIR absorption spectra clearly point to both short- and long-wavelength plasmon bands, which are attributed respectively to the transverse and longitudinal plasmon modes of the rods. Under controlled changes in the palladium-shell thickness and the refractive index of the surrounding dielectric medium, the longer-wavelength plasmon band reveals a shift, whereas the shorter-wavelength band remains invariable. Since the palladium-coated gold nanorods have an inherent photo-sensing ability, they should prove a very desirable material for the photo-catalysis and for the surface plasmon resonance-based sensors.

Results and Discussion: Example 15

The TEM image at a high magnification in Fig. 1 displays the uniform zinc oxide nanoparticles in the spherical shape.

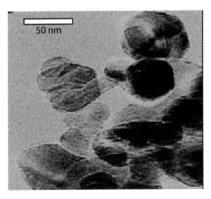

Figure 1   TEM image of the zinc oxide nanoparticles at a high magnification. Distributions of the zinc oxide nanoparticles are estimated and shown in Fig. 2.

The average diameter of the zinc oxide nanoparticles is estimated to be about 45–60 nm.

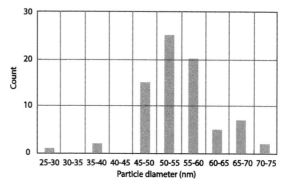

Figure 2   Distribution diagram of the zinc oxide nanoparticles.

The X-ray diffraction (XRD) pattern (not shown) consists of a dominant peak at $2\theta = 35°$. This confirms a characteristic of the hexagonal phase zinc oxide structures which is in a good agreement with the JCPDS # 36–1451 [30].

> Compare the findings with other study.

Figure 3 shows the typical optical absorption spectra of the zinc oxide nanoparticles with the different ablation times of 10, 15, and 20 min.

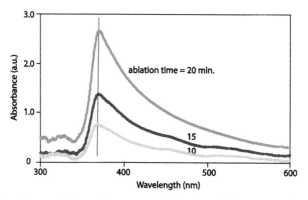

Figure 3   Measured absorption spectra of the zinc oxide nanoparticles as with various ablation times.

The spectra have a sharp peak of absorbance of all nanoparticles centered at 375 nm, corresponding to a band gap of the zinc oxide (3.3 electron volts) [31,32]. Because the zinc oxide nanoparticles prepared in the present conditions have nearly spherical shapes as seen in Fig. 1, the polarization dependence of the resonance at

> Compare the findings with other studies.

3.3 eV is also not observable in Fig. 3. Moreover, the prominent single peak is consistent with the fact that the nanoparticles present in the solution are almost spherical in shape. The broadening of the peak in Fig. 3 also indicates a band gap distribution due to various particle sizes. On the other hand, the band gaps are independent of the ablation time. This identical band gap irrespective of the ablation time indicates that the sizes of the zinc oxide nanoparticles are barely affected by the ablation time. By contrast, the absorbance of the zinc oxide nanoparticles tends to increase with the ablation time. Namely, the absorbance of 0.8, 1.5, and 2.8 corresponds to the ablation time of 10, 15, and 20 min. Since the sizes of the zinc oxide nanoparticles are independent of the ablation time, the change in the absorbance indicates that the density of the nanoparticles increases with the ablation time.

Now we discuss one possible mechanism of a generation of the zinc oxide nanoparticles in the liquid environment. That is, when the laser pulse with a high energy hits on the surface of the zinc metal plate, it vaporizes the zinc atoms, reacting with the dissolved oxygen in the aqueous solution of the sodium dodecyl sulfate containing water in order to form the zinc oxide nanoparticles. Large numbers of the zinc oxide nanoparticles arrange themselves in a specific geometry and they are covered by a layer of the sodium dodecyl sulfate in order to prevent them from further coalescence and agglomeration and to limit the growth of the particles. This offers the novel methods to produce the metallic oxide nanoparticles by further adding the amount of dissolved oxygen into the solution. In order to avoid the dissolved oxygen, one may fill the inert gas, such as argon, into the solution in order to form the pure metallic nanoparticles.

On the other hand, since the ferrous ion has visible absorption property and its absorption is dependent on the concentration of the ferrous ion, we can introduce the ferrous ion served as a probe in order to check the potential activity of the zinc oxide nanoparticles on the ferrous ion chelating to the cells. Once the MCF-7 cells are incubated with a variety of the concentrations of the zinc oxide nanoparticles under the ultraviolet irradiation for 24 h, the solutions treated with the ferrous ion are detected by absorption study. The absorption peak is about 535 nm as shown in Fig. 4.

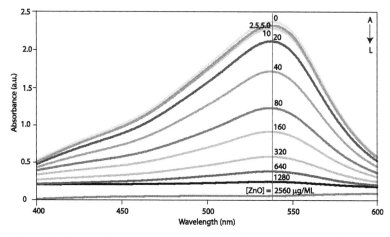

Figure 4 UV–Vis absorption spectroscopy results after treating the MCF-7 cells with different concentrations of the zinc oxide nanoparticles under the ultraviolet irradiation for 24 h.

A → L (or A, B, C, D, E, F, G, H, I, J, K, L) indicate the zinc oxide nanoparticles with concentration of (0 = the controlled cells, 2.5, 5, 10, 20, 40, 80, 160, 320, 640, 1280, 2560 μg/mL, respectively.

In Fig. 4, curve "A" represents the absorbance of the controlled cells (zinc oxide nanoparticles = 0 μg/mL) under the ultraviolet irradiation for 24 h. When the tested cells are incubated with increasing concentrations of the zinc oxide nanoparticles from 2.5 μg/mL to 2560 μg/mL under the ultraviolet irradiation for 24 h, the obvious decreasing absorbance show in curve "B" to "L", respectively, indicating the least amount of zinc oxide nanoparticles remained outside the cells. Similarly, the relation between ferrous ion chelating and concentration of the zinc oxide nanoparticles is also displayed in Fig. 5.

The size of nanoparticles, which is comparable to naturally occurring biological molecules, is a unique feature that makes them well suited for the biological applications. The nano-sized zinc oxide within a range of 10–150 nm allows its internalization into the cells, and allows it to interact with biomolecules within or on the cells, enabling it to affect the cellular responses in a selective manner [33]. Then, the zinc oxide nanoparticles can efficiently improve the permeation of the cell membrane and are diffused into the cancer cells. Consequently, the photocatalytic attacks inside the MCF-7 cells are possible.

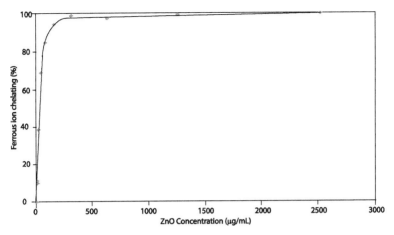

Figure 5 Ferrous ion chelating with respect to the zinc oxide nanoparticles concentration.

The assessment of the normal or dead cells is dependent on the morphological characterization. The normal cell (like a smooth cell as seen in Fig. 6 (left)) and dead cell (like a condensed or fragmented chromatin as seen in Fig. 6 (right)) are very easily distinguished. The cell-killing approach is not dominant in the controlled cells.

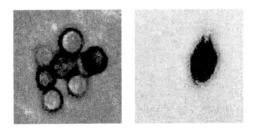

Figure 6 Living cells (left) and dead cell (right) of the MCF-7 cells.

In Fig. 7, the survival fraction of the MCF-7 cells is then plotted with respect to a variety of the concentrations of the zinc oxide nanoparticles in the presence of ultraviolet irradiation for 24 h. The lethality of the MCF-7 cells is visibly dependent on the amount of the zinc oxide nanoparticles. The cell survival of the MCF-7 cells treated by the zinc oxide nanoparticles tends to decrease when increasing concentrations of the zinc oxide nanoparticles up to 2560 μg/mL. The higher the dosage the greater the cell mortality, showing a dose-dependent effect.

This indicates that under the ultraviolet irradiation, the amount of zinc oxide nanoparticles can cause more amounts of killing the MCF-7 cells.

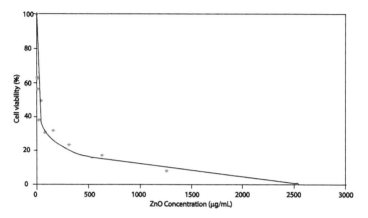

Figure 7 The cell viability of the MCF-7 cells incubated with different concentrations of the zinc oxide nanoparticles under the ultraviolet irradiation for 24 h.

Now we discuss one possible explanation of improvement in the anti-cancer activity by the zinc oxide nanoparticles under the ultraviolet irradiation. The band gap energy of the crystalline zinc oxide calculated according to the Mott model is 3.32 eV at ambient conditions. Ultraviolet irradiation contains sufficient energy to introduce the electrons to the conduction band and remain behind the holes in the valence band. These electron/hole pairs are able to enhance a series of photochemical reactions in the zinc oxide aqueous environment, generating the reactive oxygen species. At the surface of the excited zinc oxide nanoparticles, the holes can interact with hydroxyl ions or water in order to produce the hydroxyl radicals. The electrons can interact with oxygen in order to generate the superoxide radial anion. Therefore, we conclude that the zinc oxide nanoparticles under the ultraviolet irradiation possibly enhance a formation of the reactive oxygen species. The generated reactive oxygen species causes the damage of DNA and possibly lead extensive cytotoxic membrane to damage via the lipid peroxidation or protein denaturation. They finally can promote the mortality of the MCF-7 cells. These results are in agreement with the suggestion of Toduka et al. that the zinc oxide nanoparticles have the unique phototoxic effect upon the irradiation, generating cytotoxic reactive oxygen species in aqueous media. To further clarify the

Give your own suggestions

contribution of these effects, we are now investigating the ultraviolet irradiation induced enhancement in the other functional metal oxide nanostructures exposure against the human breast cancer cells. The detailed results will be presented in the next paper.

Results and Discussion: Example 16

Fig. 1 illustrates only some of many examples of the detailed fluorescence spectra for the dyes obtained in the presence (or absence) of the metallic nanostructures, Au@Pd NRs, Au NBPs, and Pd PNCs, which have different shapes. As seen from Fig. 1, the fluorescence signal from the dyes can be enhanced by adding the metallic nanostructures. This is accompanied by virtually no changes in the spectral shapes. The exception is the system denoted in concise form as Rh6G + Au@Pd NRs, for which the spectral shape becomes somewhat different from that of the pure dye. Namely, the spectral maximum of the fluorescence intensity for the latter system reveals a clear blue shift. Perhaps, the reason lies in collisions of Rh6G with Au@Pd NRs, which can change the local dielectric environment of the nanorods. The enhancement of the emission observed by us is explained by the two reasons. The first is an increased absorption rate of the dyes due to plasmonic coupling with the metallic nanostructures. The second is a possibility of orientation of the metallic nanostructures, which controls the plasmonic interactions between the metallic nanostructures and the dye molecules.

In order to examine quantitatively the influence of concentration of the metallic nanostructures on the emission intensities measured for our dyes, we have calculated the ratio of the peak fluorescence intensity for the dye–nanoparticles mixture and the relevant peak intensity for the pure dye. The values of this parameter obtained for different samples under test are displayed in Fig. 2.

The dye molecules are located very close to the metallic nanostructures at high concentrations of the metals and the dyes (e.g., for the samples A1, B1, and C1). Then the nonradiative transfer of charge carriers to the metallic nanostructures could also be significant, thus leading to decreasing fluorescence intensity. The fluorescence intensity becomes higher when the metal and dye concentrations decrease (e.g., for the samples A2, B2, and C2). This seems to be natural because the spacing between the dye molecules and the nanostructures increases and only a small number of dye molecules can couple with the metallic nanostructures. Another reason comes from weakening of the non-radiative transfer mechanism.

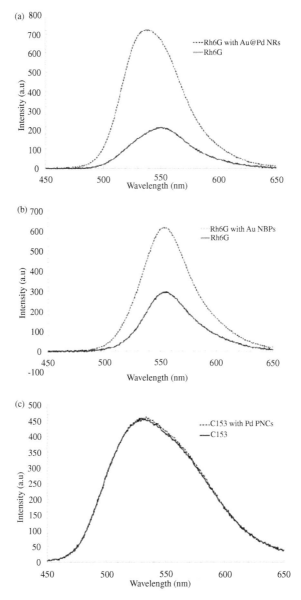

Figure 1 Fluorescence spectra of dyes (Rh6G or C153) with and without metallic nanostructures (Au@Pd NR, Au NBP, or Pd PNC). (a) Fluorescence intensity of Rh6G in the presence and absence of Au@Pd nanorods (Sample A3). (b) Fluorescence intensity of Rh6G in the presence and absence of Au nanobipyramids (Sample B2). (c) Fluorescence intensity of C153 in the presence and absence of single-crystalline porous Pd nanocrystals (Sample C2).

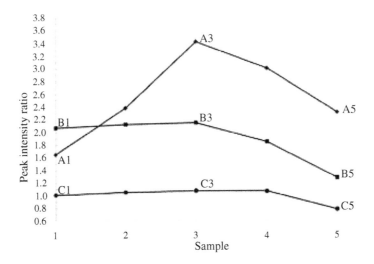

Figure 2 Fluorescence enhancement of dyes (Rh6G or C153) for sample set A, B, and C.

There exists an optimum concentration that provides efficient plasmonic interactions for each set of our samples (e.g., for the cases referred to as A3, B3, and C3). Under these optimum conditions, the fluorescence intensity is enhanced approximately by 3.5 times for the system Rh6G + Au@Pd NRs, by 2.2 times for Rh6G + Au NBPs, and by 1.1 times for C153 + Pd PNCs. Beyond this optimum (see, e.g., the data for the samples A5, B5, and C5), the fluorescence intensity tends to decrease again with decreasing concentrations of the dye–nanoparticles mixtures. Probably, this can arise from re-quenching of the intrinsic fluorescence originated from mismatched concentrations of the solutions. This finding should be further confirmed by the theoretical calculations based upon the finite-difference time-domain technique. The latter will be a subject of our forthcoming study.

Give your own suggestions

6.2 Problems

Pb.1. Discuss how to get good results and write proper discussion.

Pb.2. Solve the problems when your results and discussion are not satisfied by your advisor.

Pb.3. Write some examples of tables, figures, and equations.

Pb.4. Suppose you are a reviewer; give some comments on the following papers:

Results and Discussion (A)

Figure 1 displays the azimuthal angle dependence of the SH intensity for 19 nm width × 40 nm periodicity Cu nanowires. The SH intensity depends strongly on the sample rotation angle ϕ and the polarization combinations. We see two main lobes in p-in/p-out, s-in/p-out, and s-in/s-out polarization configurations in Figs. 1(a), (c), and (d), respectively. The SH intensity is below the noise level for p-in/s-out polarization configuration in Fig. 1(b).

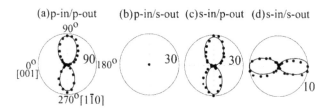

Figure 1 Measured (filled circles) and calculated (solid lines) SH intensity patterns for 19 nm width × 40 nm periodicity Cu nanowires as a function of the sample rotation angle ϕ for the fundamental photon energy of 1.17eV. The rotation angle ϕ is defined as the angle between the incident plane and the wire axes of the [001] direction. The incident angle is 45°. The four different input and output polarization combinations are indicated at the top. Rotational anisotropy in SHG is plotted in polar coordinates, with 0° corresponding to the [001] substrate direction, and 270° corresponding to the [1$\bar{1}$0] substrate direction. The maximum of the intensity in a relative scale is written beside each pattern.

We have analyzed the origin of the SHG from Cu nanowires shown in Fig. 1 (filled circles) by a phenomenological model [5]. We have fitted theoretical SH intensity patterns to the experiment by a least square fitting program based on the Cs symmetry of the Cu nanowires/ NaCl(110) system. The contributions of $\chi^2_{222}, \chi^2_{311}, \chi^2_{323}$ elements are found to be dominant, and they reproduce the experiment well (solid lines in Fig. 1). The indices 1, 2, and 3 denote the [001], [1$\bar{1}$0], and [110] directions on the NaCl(110) substrate, respectively. The χ^2_{222}

element arises from the symmetry breaking by the present ridge-to-valley structure as self-organization templates, while we suggest that the SH response from the electromagnetic field normal to the substrate plane enhances the contributions $\chi_{311}^2, \chi_{323}^2$.

In order to clarify the latter point, numerical computation for the electric field contribution close to Cu nanowires is performed by using finite-different time-domain (FDTD) method [6]. In Fig. 2(a), we assume that the wires are of infinite length and circular cross-section. We model the Cu nanowires by four arrays of cylinders, and they are infinitely extended along the direction 1. The plane waves polarized perpendicular to the cylinder axes or in direction 2 move in direction 3 and pass through the cylinders. We use Drude–Lorentz parameters fit to empirical dielectric constant data of Cu obtained from literature [7] in the photon energy of interest. We consider a two-dimensional FDTD unit cell corresponding to the size of 1 nm × 1 nm, and a typical propagation is over a time period of 500 fs. The intensity of the total field is normalized to that of the incident field intensity. Figure 2(b) shows the fundamental field intensity at the observation point shown in Fig. 2(a) as a function of the incident photon energy obtained by FDTD calculations. The cylinder periodicities are from 4 to 40 nm at a constant cylinder diameter of 20 nm. The field amplitude perpendicular to the cylinder axes reaches the maximum near the photon energy of 2.2 eV. The main resonance is red-shifted slightly with decreasing separation distance. The incident field parallel to the cylinder axes is too weak to allow us for discussion of its field intensity.

Figures 3(a)–(d) and (e)–(h) demonstrate the field distribution at the single photon energy of 2.2 eV for the electric field in the 1:[001] and in the 3:[110] direction on the surface, respectively. The field near the cylinders of 20 nm diameter × 40 nm periodicity in Figs. 3(a) and 3(e), corresponding to Cu nanowires of 19 nm width × 40 nm periodicity used for SHG measurements, is about two and six times the excitation amplitude, respectively. The enhancement is much larger for cylinders with smaller gap. In Figs. 3(d) and 3(h), the electric field near the cylinders of 4 nm separation distance is almost six and thirty-six times as large as the incident field, respectively. We suggest that two distinct types of interaction effects may occur according to the increase of the periodicity. In a considerably short distance, short-range interactions between neighboring cylinders induce near-field coupling and create highly sensitive plasmons confined to metal boundaries. However, when the periodicity exceeds the range of near-field coupling, far-field interactions prevail among cylinders. This mechanism has been explained by using a dipole–dipole interaction model [8]. Dipole field in an individual cylinder induced dipoles in the neighboring cylinders by distorting the neighbor's electron cloud leads to the formation of localized surface plasmons and local field enhancement.

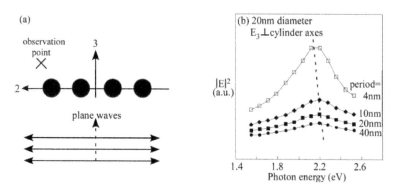

Figure 2 (a) Coordinate system of four-cylinder arrays of Cu used for FDTD calculation. The measurement point is indicated at the upper left corner near the cylinders. The incident field is perpendicular to the cylinder axes. (b) Electric field intensity calculated by FDTD simulations of Cu cylinders of 20 nm diameter with different separation distances as a function of photon energy.

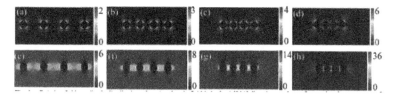

Figure 3 Relative field amplitude distribution when the electric field is in the 1:[001] direction on the surface at the photon energy of 2.2 eV for four interacting cylinders at the same diameter of 20 nm with separation distance (a) 40 nm, (b) 20 nm, (c) 10 nm, and (d) 4 nm. Relative field amplitude distribution when the electric field is in the 3:[110] direction on the surface at the photon energy of 2.2 eV for four interacting cylinders at the same diameter of 20 nm with separation distance (e) 40 nm, (f) 20 nm, (g) 10 nm, and (h) 4 nm. The amplitude of the electric field is shown in arbitrary unit and is in a linear color scale on the right of each image.

By the analysis of the theoretical calculation and the measured SH intensity patterns from Cu nanowires, we conclude that Cu nanowires can provide enhancements in the local field in the direction perpendicular to the substrate via a coupling of the localized surface plasmon resonances. These effects are responsible for the dominant contributions of $\chi^2_{311}, \chi^2_{323}$ in such nanowires.

Results and Discussion (B)

First we analyzed the field-amplitude enhancement of the nonlinear optical process represented by χ^2_{323} element taking the input field in (2, 3) direction into account. We show in Fig. 1 the FDTD calculated field distribution in direction 3 for Cu nanowires at the photon energy of 2.2 eV. For the same input field in (2, 3) direction, the enhancements of the FDTD calculated field in direction 2 are observed similarly (not shown). Cu nanowires have cross-sections of circular, square, and triangular shapes in Figs. 1(a)–(d), (e)–(h), and (i)–(l), respectively. The color scales in the images are of the relative amplitude of the electric field. Orange-to-red indicates the highest intensity and dark blue indicates the lowest intensity.

The maximum amplitudes are found to depend strongly on the separation distance. The enhancement represented by the red scales in the images around each gap between wires is much larger for the wires with smaller gaps than for those with larger gaps. For instance, the maximum electric field near the circular wires of 4-nm separation distance shown in Fig. 1(a) is about three times as intense as that of 40-nm separation distance shown in Fig. 1(d).

This enhancement is induced by the excitation of dipolar plasmon mode because it is consistent with our linear absorption spectra showing plasmon maxima near 2.4 eV when the electric field is perpendicular to the long wire axes [4]. According to Rechberger et al., two distinct types of interaction between the electric field and the nanowires should occur depending on the periodicity [9]. For the short periodicity, near-field coupling between neighboring wires is dominant due to local plasmons confined within metal boundaries. When the periodicity is over the range of the near-field coupling, dipole field induced in the wires distorts the neighboring electron cloud and leads to the formation of the localized surface plasmon waves. Therefore, we expect that the field strength changes drastically as a function of the periodicity due to the plasmon excitation. This local field enhancement may lead to a strong enhancement of the second-order nonlinear polarization through the χ^2_{323} element.

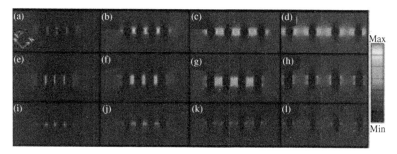

Figure 1 Calculated relative field-amplitude distribution in direction 3 when the input electric field is in (2, 3) direction at the photon energy of 2.2 eV for four circular wires at the same diameter of 20 nm with separation distance (a) 4 nm, (b) 10 nm, (c) 20 nm, and (d) 40 nm; for four square wires at the same perimeter of 72 nm with separation distance (e) 4 nm, (f) 10 nm, (g) 20 nm, and (h) 40 nm; and for four triangular wires at the same perimeter of 84 nm with separation distance (i) 4 nm, (j) 10 nm, (k) 20 nm, and (l) 40 nm with the dielectric constant of that of Cu.

Next, we analyzed the enhancement of the nonlinear optical process represented by the χ^2_{311} element taking the input field in direction 1 into consideration. Figure 2 demonstrates the FDTD calculated field distribution in direction 1 for Cu nanowires at the photon energy of 2.2 eV. The maximum amplitudes are found to depend strongly on the cross-sectional shapes of the wires. The four bright spots near each wire in Figs. 2(a)–(d) and (e)–(h) show the field-amplitude enhancement of the circular and square wires, respectively. In Figs. 2(i)–(l), two bright spots indicate the enhancement of the field near each triangular wire: The enhancement is large at the top-left and top-right corners and is very small at the lower corner.

In every case, the field peaks in strength near the corners and falls off rapidly inside the metallic nanostructures of the different shapes. The observed field strength does not change significantly as a function of the periodicity. This is consistent with the prediction by Rechberger et al. saying that there should be no interaction effect between the electric field and the nanowires created by the local plasmons for the field distribution in direction 1. Therefore, we can exclude the enhancement of the local field by the plasmon excitation from the main origins of the χ^2_{311} element.

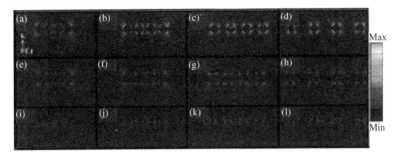

Figure 2 Calculated relative field-amplitude distribution in direction 1 when the input electric field is in direction 1 at the photon energy of 2.2 eV for four circular wires at the same diameter of 20 nm with separation distance (a) 4 nm, (b) 10 nm, (c) 20 nm, and (d) 40 nm; for four square wires at the same perimeter of 72 nm with separation distance (e) 4 nm, (f) 10 nm, (g) 20 nm, and (h) 40 nm; and for four triangular wires at the same perimeter of 84 nm with separation distance (i) 4 nm, (j) 10 nm, (k) 20 nm, and (l) 40 nm with the dielectric constant of that of Cu.

Now we discuss the lightning-rod effect as a candidate origin of the SH response. Sharp corners of nanowire structures should behave as optical dipole antennas and their strong field should couple with the incident light beam even if there is no plasmon contribution. This coupling provides sufficient field enhancement to allow for the optical nonlinearity depending on the cross-sectional shapes of the wires. These features are consistent with the results in Fig. 2. This is called lightning-rod effect and has also been observed at the corners of Ag strips and Au nanoparticles [10–12]. Hence, we strongly suggest that the lightning-rod effect created by the geometric singularity of the sharply pointed wire structures plays an important role for the dominant contribution of the χ^2_{311} element in Cu nanowires.

We have found that the maximum amplitudes for the FDTD calculated field distribution in direction 3 at the fundamental photon energy of 2.2 eV (Fig. 1) are about five times as intense as that of the field distribution in direction 1 (Fig. 2). This indicates that the enhancement of the χ^2_{323} element is dominant for the electric field components along the surface normal. At the same fundamental photon energy of 2.2 eV, second-harmonic generation (SHG) due to the χ^2_{311} element is not enhanced very much for the polarization B due to the weak input field. Nevertheless, the 311 component was observable in our experiment. We suggest that the enhancement of the χ^2_{311} element may be dominated due to the enhancement of the field at frequency 2ω. For further development of the calculation, it is necessary to consider the enhancement for the output field distributions at any frequencies from the structural materials of interest.

Results and Discussion (C)

Figure 1(a) shows a linear image of the ZnS polycrystalline sample under illumination by white light. In Fig. 1(b) the SF intensity image of the same sample is shown by using the confocal sum frequency (SF) microscopy with a pinhole of 2 mm diameter. With the objective lens of magnification 20×, the nominal lateral resolution was 100 μm. The bright spots in the figure represent the SF photons. In Fig. 1(b) the sizes of the smallest distinguished surface pits are as small as 2 μm and are much smaller than the nominal resolution. The lateral resolution is thus determined by the focusing of the laser beam at 532 nm in the present condition.

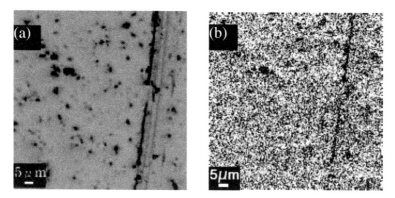

Figure 1 (a) Linear image of ZnS polycrystalline pellet under illumination by white light. (b) Confocal SF intensity image with a pinhole of 2 mm diameter. Bright spots represent photons.

According to Cox and Sheppard, the resolution of confocal microscopy is given by

$$r = \frac{0.61\lambda}{\sqrt{2}NA} \tag{1}$$

Here λ is the wavelength and NA is the numerical aperture of the objective lens [8]. When we put λ = 460 nm and NA = 0.45 for our confocal SF microscopy into Eq. (1), we obtain a resolution of 0.44 μm. This theoretical spatial resolution is better than that of our previous conventional one [4]. The size of the pinhole must be 8.8 μm in diameter in order to achieve this theoretical resolution. However, due to the starting trial demonstration of this method, the current resolution of the confocal SF microscope is far worse than this Fourier limit. Still we have succeeded in the construction of the first optical confocal SF microscope and obtained confocal microscopic SF intensity images by using ZnS polycrystals as a sample. It was also checked to reject out-of-focus light rays.

A confocal SF microscopy with a pinhole of 0.4 mm diameter allowed us to obtain a narrower depth of the field distribution so that it can form three-dimensional images of thick objects. With the objective lens of magnification 20×, the lateral resolution should nominally be 20 μm. The ideal resolution in the depth of the confocal microscope is given by [9]

$$Z = \frac{1.4n\lambda}{NA^2} \tag{2}$$

n is the refractive index of the medium. When we put the refractive index n of ZnS = 2.29 at λ = 460 nm and NA = 0.45 into Eq. (2), we obtain Z = 7.3 μm. However, due to the larger pinhole than that of the ideal size, the resolution in the depth is estimated approximately to be 21 μm.

The demonstration of the three-dimensional scan was performed by taking the optical images of ZnS planes of different depths as shown in Fig. 2. The image of the sample under illumination by white light is shown in Fig. 2(a). In Fig. 2(b), (c), and (d) the two-dimensional SF intensity distribution at three different depths 0, 5, and 10 μm, respectively, is shown in three-dimensional displays with the intensity in the vertical direction, in order to highlight the difference.

The presence of a highly defective region was observed in the side illumination image (not shown) and resulted in the weak SF intensity near the left edge of the sample in Fig. 2(b). We find contrast in the SF intensity distribution when the incident beams were focused at different depths in the sample. The edge curve of the sample seen in the SF intensity image in Fig. 2(c) is rather straight, while those in Figs. 2(b) and (d) are more like that in the linear image in Fig. 2(a). The distributions of defects in the ZnS polycrystals are interpreted to be different between these depths.

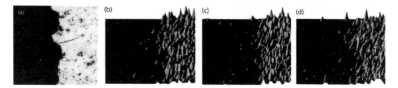

Figure 2   (a) Linear image of ZnS under illumination by white light with an area of 100 × 100 μm². Confocal SF intensity image of ZnS with a pinhole of 0.4 mm diameter at (b) 0 μm, (c) 5 μm, and (d) 10 μm depths. The size of a scanned area of (b), (c), and (d) was 100 × 100 μm². The thickness of the ZnS sample was 3 mm. The photon energy of the incident IR light was 2890 cm⁻¹.

The SF intensity at the sample surface obtained from a confocal pinhole of 0.4 mm diameter in Figs. 2(b)–(d) was found to be much weaker than that of 2 mm diameter in Fig. 1(b). There are two ways to increase the SF intensity. This can be performed by averaging data from many frames. This has the drawback of slowing down the effective frame rate of the microscope. Alternatively, it can be carried out by raising the intensity of the excitation light up to the highest limit below the damage threshold of the specimen. The detailed results will appear in our next paper.

Results and Discussion (D)

TEM images in Figs. 1(a), (b), and (c) showed the uniform Au, Cu, and Pt nanowires, respectively, as long and straight periodic dark images parallel to the [001] direction of the substrates. The minimum widths of the nanowires were 40 nm for Au, 14 nm for Cu, and 9 nm for Pt. One of the key parameters determining the minimum width of the nanowires deposited onto a crystalline substrate is the surface energy of the wire metal. High surface energy of metals will lead to high contact angle on the substrate surface [7]. As the surface energy is the largest for Pt, the second largest for Cu, and the smallest for Au in their liquid state, the minimum width of Cu nanowires (14 nm) is larger than that of Pt (9 nm), but is smaller than that of Au (40 nm).

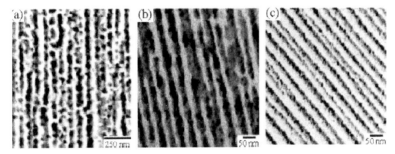

Figure 1 TEM images of Au (a), Cu (b), and Pt (c) nanowires.

Next we measured the SH intensity spectra from the obtained metallic nanowires as a function of the SH photon energy for the p-in/p-out polarization configuration as shown in Fig. 2. The incident wave vector of the excitation was parallel to the [110] direction of the NaCl (110) faceted substrates. The SH intensity spectra taken from Au nanowires in Fig. 2(a) showed a nearly flat dependence of SH intensity on the photon energy below $2\hbar\omega$ = 3.0 eV. The intensity abruptly increased above $2\hbar\omega \sim$ 3.1 eV, and reached a maximum at 3.3 eV. The SH intensity spectra for Cu nanowires in Fig. 2(b) showed steady increase above $2\hbar\omega$ = 3.0 eV. The SH response exhibited a peaked resonance near 4.4 eV. In Fig. 2(c) the SH intensity spectra for Pt nanowires showed steady increase above $2\hbar\omega$ = 3.7 eV. The SH response exhibited a peaked resonance near 4.3 eV. The main peak near $2\hbar\omega$ = 3.3 eV for Au nanowires in Fig. 2(a) appeared to have a similar origin as the main peak at 4.4 eV for Cu nanowires in Fig. 2(b), and the main peak at 4.3 eV for Pt nanowires in Fig. 2(c).

We performed a theoretical analysis of the SH intensity in order to determine the values of the local field factor and then to find whether the surface plasmon resonance served as an origin of the enhancement of the SH response from the metallic nanowires. We calculated the local field enhancement factor for Au, Cu, and Pt nanowires by a quasi-static theory [8]. It was found that the calculated spectra for the metallic nanowires in Fig. 3 reproduced the experimental findings in Fig. 2 well. Hence, these theoretical results indicated that the resonant enhancement near the SH photon energy of 3.3 eV for Au, 4.4 eV for Cu, and 4.3 eV for Pt originated from the coupling of plasmon resonances in the metal nanowires. Our experimental findings and calculations were consistent with the suggestion by Schider et al. that the plasmon excitation in the metallic nanostructures gave rise to the local electric field enhancement [2]. For Au nanowires, it was noteworthy that the nonlinear response was weak at the SH photon energy lower than 3.3 eV, due to the anisotropic depolarization field [3].

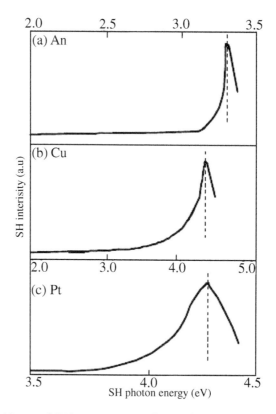

Figure 2 Measured SH intensity spectra from different metallic nanowires as a function of the SH photon energy for the p-polarized input and p-polarized output combination. The solid lines were guide to the eyes.

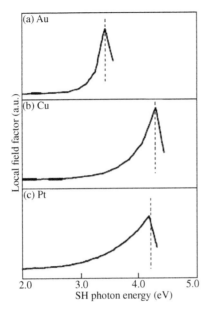

(a) Au

(b) Cu

(c) Pt

Local field factor (a.u.)

SH photon energy (eV)

2.0   3.0   4.0   5.0

Figure 3   Calculated SHG local field enhancement factor as a function of the SH photon energy for different metallic nanowires based on Ref. [8].

Results and Discussion (E)

TEM images showed uniform Cu nanowires as long and straight periodic dark images parallel to the [001] direction of the substrates. Minimum width of nanowires was 14 nm [3]. We then measured azimuthal angle dependence of SH intensity from the obtained Cu nanowires as shown in Fig. 1. The SH intensity from nanowires depended strongly on the sample rotation angle ϕ and the polarization configurations. The SH intensity patterns from Cu nanowires displayed two main lobes for the p-in/p-out, s-in/p-out, and s-in/s-out polarization configurations in Figs. 1(a), (c), and (d), respectively.

We analyzed SH intensity patterns from these nanowires in Fig. 1 (filled circles) by a phenomenological model [7]. The fitted results were shown in the solid curves in Fig. 1. Contributions of the $\chi^2_{323}, \chi^2_{311}, \chi^2_{222}$ elements were proved to be prominent the SH signal for Cu nanowires. The indices 1, 2, and 3 denote the [001], [1$\bar{1}$0], and [110] directions on the NaCl (110) surface, respectively. The pattern for Cu nanowires for the s-in/s-out polarization configuration in Fig. 1(d) was seen to be dominated by the contribution of the χ_{222}^{2} element. This element resulted from the broken symmetry in the 2:[1$\bar{1}$0] direction due to the cross-sectional shapes of the nanowires produced by the periodic macrosteps parallel to the 1:[001] direction.

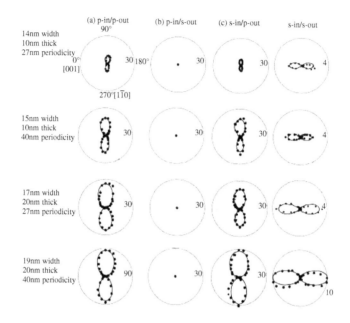

Figure 1 Measured (filled circles) and calculated (solid line) SH intensity patterns for different wire widths of Cu nanowires as a function of the sample rotation angle ϕ at the fundamental photon energy of 1.17 eV. The angle ϕ is defined as the angle between the incident plane and the [001] direction on the sample surface. The incident angle is 45°. The SH intensity is plotted in polar coordinates. The intensity in a relative unit is written next to each pattern. The four different input and output polarization combinations are indicated at the top.

We numerically investigated the local electromagnetic field by FDTD method to clarify contributions of the χ_{323}^{2}, χ_{311}^{2} elements dominated the optical nonlinearity of Cu nanowires [3]. It was found that contribution of the χ_{323}^{2} element originating from local plasmons in nanowires dominated the optical nonlinearity for the p-in/p-out polarization configuration in Fig. 1(a), and the contribution of the χ_{311}^{2} element originating from the lightning-rod effect dominated the optical nonlinearity for the s-in/p-out polarization configuration in Fig. 1(c) [7]. From the analysis of the SH intensity patterns by a phenomenological model and FDTD method, we have concluded that the electric field component along the surface normal enhanced by the plasmon-resonant coupling in the wires was observed for Cu nanowires.

Next we measured SH intensity spectra from nanowires as a function of SH photon energy as shown in Fig. 2. The SH intensity spectra for Cu nanowires showed steady increase above $2\hbar\omega = 3.0$ eV. The SH response exhibited a peaked resonance near 4.4 eV. We considered the enhancement of the local field by the plasmon excitation in the nanowires as candidate origin of the SHG from Cu nanowires.

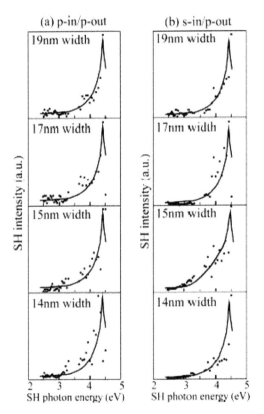

Figure 2 Measured SH intensity spectra from Cu nanowires of different wire widths as a function of the SH photon energy for (a) p-in/p-out and (b) s-in/p-out polarization configurations. The incident wave vector of excitation is parallel to the [1$\bar{1}$0] direction of the NaCl(110) faceted substrates. The solid lines are guide to the eyes.

We performed a theoretical analysis of SH intensity to determine values of the local field factor and then to find whether surface plasmon resonance served as the origin of enhancement of the SH response from nanowires. We calculated local field enhancement factor for Cu nanowires by a quasi-static theory [8]. The calculated enhancement factor was described in [3,9]. It was found that the calculated spectra for the nanowires reproduced the experimental findings in Fig. 2 well [10]. Hence, these theoretical results indicated that the resonant enhancement near the SH photon energy of 4.4 eV originated from the coupling of plasmon resonances in the metal nanowires. Our experimental findings and calculations were consistent with the suggestion by Schider et al. that the plasmon excitation in the metallic nanostructures gave rise to the local electric field enhancement [2].

Results and Discussion (F)

Results

TEM images in Fig. 1 show uniform Cu nanowires as long and straight periodic dark images parallel to the [001] direction of the substrates. At the nominal thickness of 5 nm, Cu forms arrays of nanodots (Figs. 1e and f). The arrays of nanowires are formed at the nominal thickness of 10 nm, and the minimum widths of 14 nm are obtained (Fig. 1b). Cu nanodots fabricated in the present work have spherical shapes as seen in Fig. 1f and thus the polarization dependence of the plasmon resonance at 2.25 eV is not observable as is seen in Fig. 2. These absorption spectra are in agreement with the strong absorption band at about 2.21 eV of spherical Cu nanoparticles in the literatures [12,13]. This result for nanodots serves as a control experiment for the results for nanowires in Fig. 3. In the absorption spectra of Cu nanowires, we observe absorption maxima near 2.25 eV for perpendicular polarization in Fig. 3b and they are red-shifted for the parallel polarization in Fig. 3a. We also find the red-shift of the absorption maxima when the widths of nanowires are increased. Our measurements of the energy shifts of the absorption maxima were quite reproducible among several runs at different positions from each sample within the coefficient of variation of less than 5%.

Discussion

One of the key parameters determining minimum width of nanowires deposited onto a crystalline substrate is surface energy of wire metal.

High surface energy of metals will lead to high contact angle on substrate surface. As surface energy is the largest for Pt, the second largest for Cu, and the smallest for Au in their liquid state [14], the minimum width of Cu nanowires (14 nm) is larger than that of Pt (9 nm) [15], but is smaller than that of Au (40 nm) [8].

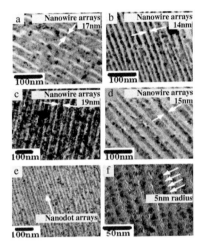

Figure 1 TEM images of Cu nanowires of samples (a) C7, (b) C8, (c) C9, and (d) C10, and TEM images of Cu nanodots of sample C12 at (e) low magnification and (f) high magnification.

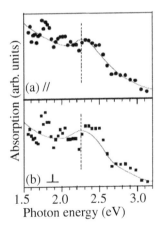

Figure 2 Absorption spectra of 10 nm Cu nanodots on the faceted NaCl(110) template. // (⊥) denotes the electric field parallel (perpendicular) to the wire axes. Solid line is guide to the eyes.

Figure 3 shows that blue-shift of the absorption maxima occurs with the small wire widths. Since the wire dimensions and separation distances are relatively small compared to the wavelength of incident light, we have applied Maxwell–Garnett theory to address this size effect on absorption spectra [16]. This theory defines an effective dielectric constant for a composite from the dielectric constants of constituent metal and the host medium.

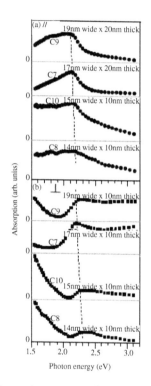

Figure 3 Measured absorption spectra of Cu nanowires sandwiched by SiO layers. // (⊥) denotes the electric field parallel (perpendicular) to the wire axes.

By assuming metal volume fraction of 0.1 and various aspect ratios a/b (a and b are defined as long and short wire axis lengths, respectively), we calculate their absorption spectra as shown in Fig. 4. Absorption maxima for incident field polarized along the wire axes occur at lower photon energy than those for incident field polarized perpendicular to the wire axes. This is consistent with experimental result, but we find that detailed experimental dependence of energy positions of absorption maxima on polarization is different from prediction by Maxwell–Garnett model. Since Maxwell–Garnett model cannot reproduce systematic changes in polarization dependences of absorption intensity qualitatively, we suggest that this model is not appropriate for predicting our absorption spectra. In order to remove the first and second sources of discrepancies mentioned above, rigorous numerical studies using finite-difference time-domain method of solving electromagnetic problems by integrating Maxwell's differential equations are under way. This method will also allow us to consider actual wire size dependence of peak energy positions.

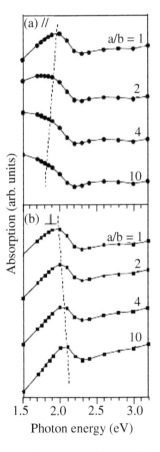

Figure 4 Calculated absorption spectra of Cu nanowires at different aspect ratio a/b. a (b) is the long (short) wire axes. // (⊥) denotes the electric field parallel (perpendicular) to the wire axes.

Results and Discussion (G)

Figures 1, 2, and 3 display fluorescence intensity as a function of time with the 2 mm, 4 mm, and 6 mm ICG targets in the phantom, respectively. For all targets, the measured fluorescence intensity is found to be the largest for 90° detection point because the target is nearest to the detection point as shown in Fig. 1. The intensity becomes weaker when changing the detection point from 90° to 30° or 180° because the target is away from the detection point.

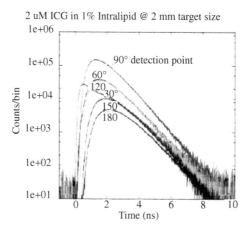

Figure 1   Fluorescence temporal profiles for 2 mm ICG target.

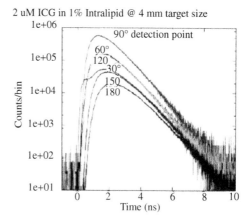

Figure 2   Fluorescence temporal profiles for 4 mm ICG target.

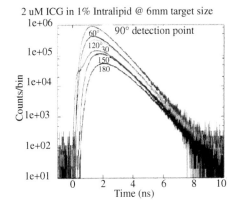

2 uM ICG in 1% Intralipid @ 6mm target size

Figure 3 Fluorescence temporal profiles for 6 mm ICG target.

On the other hand, a contamination of the excitation light is found at the point close to the excitation because the excitation light intensity was very high and eventually a small amount of excitation light was also detected by the system. This causes a hump for all targets. However, the amplitude of the hump is reduced when a target size is increased to be 6 mm as shown in Fig. 3. This is due to suppression of contaminations by high amount of fluorescence photons from the target.

Now we analyze the temporal profile. The profiles at 30° (60°) and 150° (120°) show an initially large difference and then merge each other beyond about 2 ns. The coincidence beyond 2 ns can be explained by the geometrical symmetry as shown in Fig. 1.

Disagreement less than 1 ns can be explained by a peak intensity of the temporal profile, a mean transit time, and a standard deviation. The ratios of the peak intensity of the temporal profile between 60° and 120° with the 2 mm, 4 mm, and 6 mm targets are 1.07, 1.10, and 1.29, respectively. The ratios are larger when the targets become bigger. This is because the asymmetry of the data is larger with larger target size.

A mean transit times (MTTs) of the temporal profile at 60° (120°) with the 2 mm, 4 mm, and 6 mm targets are 2.06 ns (2.14 ns), 2.08 ns (2.14 ns), and 1.96 ns (2.09 ns), respectively. On the other hand, standard deviations (SD) of the profile at 60° (120°) with the 2 mm, 4 mm, and 6 mm targets are 1.03 ns (1.02 ns), 1.00 ns (1.01 ns), and 0.97 ns (0.98 ns), respectively. The target size dependence of MTT tends to be similar to the SD. The MTT and SD are smaller when the targets become bigger. This change is attributed to high contribution of absorption of the excitation light with larger target size. Therefore, we suggest that the target size plays a key role for the temporal profile change in an early part, but not in a later part. This geometry contribution is expected to be a key parameter to obtain more precise reconstruction image.

Results and Discussion (H)

We analyzed the second harmonic intensity patterns from the nanowires in Fig. 1 (filled circles) by a phenomenological model [11]. The fitted results were shown in the solid curves in Fig. 1. The rotation–angle dependences of the second-order susceptibility tensor components for different metallic nanowires were indicated in Table 1. The numbers 1, 2, and 3 (Fig. 1) of the tensor elements represented the [001], [1$\bar{1}$0], and [110] template directions, respectively. The first suffix (i) of the χ^2_{ijk} tensor elements showed the output field distribution, whilst the remaining suffix (jk) meant input electric field component.

Table 1 Rotation-angle dependences of the second-order susceptibility tensor components for different metallic nanowires

Nanowires	p-in/p-out	p-in/s-out	s-in/p-out	s-in/s-out
Au	$\chi^{(2)}_{113}, \chi^{(2)}_{223}, \chi^{(2)}_{333}$	$\chi^{(2)}_{113}, \chi^{(2)}_{223}$	$\chi^{(2)}_{311}, \chi^{(2)}_{333}$	
Cu	$\chi^{(2)}_{323}$		$\chi^{(2)}_{311}$	$\chi^{(2)}_{222}$
Pt	$\chi^{(2)}_{113}, \chi^{(2)}_{223}, \chi^{(2)}_{323}$	$\chi^{(2)}_{113}, \chi^{(2)}_{223}$	$\chi^{(2)}_{311}$	$\chi^{(2)}_{222}$

The $\chi^2_{113}, \chi^2_{223}$ elements dominated second-order optical nonlinearity of Au nanowires for the p-in/s-out polarization configuration in Fig. 1(b) partly due to a strong depolarization field in the wires. This similar effect was observed for Pt nanowires in Fig. 1(j). We suggested that the suppression of the $\chi^2_{113}, \chi^2_{223}$ elements due to a weak depolarization field in Cu nanowires led to the intensity of second harmonic generation as low as the noise level for the p-in/s-out polarization configuration in Fig. 1(f).

The prominent $\chi^2_{113}, \chi^2_{223}$ elements also led to an enhancement of the optical nonlinearity of Au and Pt nanowires for the p-in/p-out polarization configuration. As clearly seen in this polarization configuration, the depolarization effect from Au nanowires in Fig. 1(a) could be judged to be stronger than that from Pt nanowires in Fig. 1(i). The two-lobed patterns in Au nanowires were wider in the middle in p-in/p-out (Fig. 1(a)) and s-in/p-out (Fig. 1(c)) polarization configurations due to the contribution from the χ^2_{333} element.

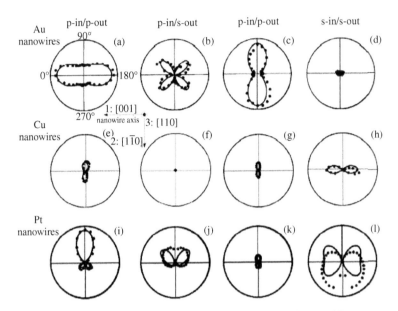

Figure 1 Measured (filled circles) and calculated (solid line) second harmonic intensity patterns for different metallic nanowires as a function of the sample rotation angle ϕ.

The pattern for Cu nanowires for the s-in/s-out polarization configuration in Fig. 1(h) was seen to be dominated by the contribution of the χ^2_{222} element. This element resulted from the broken symmetry in the 2:[1$\bar{1}$0] direction due to the cross-sectional shapes of the nanowires produced by the periodic macrosteps parallel to the 1:[001] direction. The similar effect could be observed for Pt nanowires in Fig. 1(l); however, it was not observed for Au nanowires in Fig. 1(d) because the incident electric field in the 2:[1$\bar{1}$0] direction was partly canceled by the depolarization field perpendicular to the wire axes. This difference was also suggested to be due to the larger anisotropy in the nonlinearity of Cu and Pt nanowires due to thinner wires.

We numerically investigated the local electromagnetic field by the finite-difference time-domain method in order to clarify the contributions of the $\chi^2_{311}, \chi^2_{323}$ elements dominated the optical nonlinearity of Cu nanowires [1]. It was found that the contribution of the χ^2_{323} element originating from local plasmons in nanowires dominated the optical nonlinearity for the p-in/p-out polarization configuration in Fig. 1(e), and the contribution of the χ^2_{311} element originating from the lightning-rod effect dominated the optical nonlinearity for the s-in/p-out polarization configuration in Fig. 1(g). These similar effects also led to an enhancement of the optical nonlinearity of Pt nanowires. The contribution of χ^2_{323} element in Pt nanowires originating from the plasmon resonances dominated further optical nonlinearity for the p-in/p-out polarization configuration in Fig. 1(i), and the contribution of the χ^2_{311} element originating from the lightning-rod effect dominated the optical nonlinearity for the s-in/p-out polarization configuration in Fig. 1(k). For Au nanowires, it was possible that the contribution of the χ^2_{311} element originating from the lightning-rod effect dominated the optical nonlinearity for the s-in/p-out polarization configuration in Fig. 1(c). The contribution of the χ^2_{323} element originating from the plasmon resonances was not observed for the p-in/p-out polarization configuration in Fig. 1(a) due to a strong depolarization field in Au nanowires.

It was noted that the calculated and measured second harmonic intensity patterns for p-in/s-out and s-in/s-out of the Pt nanowires were different. This fact indicated that the simple modeling of the Pt nanowire arrays as flat thin layers was not appropriate and the geometrical effect of the Pt nanowires and the template were not optically negligible. These problems could be solved if the incidence angle dependence of the SH intensity was also considered [12]; however, it was not available due to the short life time of our Pt samples under the laser irradiation.

Chapter 7

Research Ethics

After you finish writing the paper, checking plagiarism is necessary before you submit it for publication. In this chapter, I will briefly introduce ethics in research. Research ethics are based on the standards set by the Committee on Publication Ethic (COPE). All papers not conforming to these standards are removed from publication if plagiarism or duplicate content is found, even after the publication. Therefore, all parties involved—participants and researchers—are expected to follow ethical behavior in accordance with these standards.

7.1 Ethics

Ethics concerning participants

1. You must inform the participants about the nature of your study and get their consent. Research with children below 18 years needs their parents' consent.
2. You must inform the participants that their participation is voluntary. They can withdraw anytime without giving reasons.
3. You should tell them that all information acquired from them would be kept safe and would be accessed by the researcher only.

Research Methodologies for Beginners
Kitsakorn Locharoenrat
Copyright © 2017 Pan Stanford Publishing Pte. Ltd.
ISBN 978-981-4745-39-0 (Hardcover), 978-1-315-36456-8 (eBook)
www.panstanford.com

4. You must not keep anything secret from the participants as far as possible. If something cannot be revealed immediately, tell the truth as soon as the study ends.

Ethics concerning researchers

1. You must be open about the duties and authorship. It is important to put them on paper and get the paper signed.
2. The authorship must be made clear at the beginning of the research. You should better put them on paper and get the signature of all parties.
3. You must inform who can use the data and claim the authorship for a future paper on the same data. You must get an agreement signed.

7.2 Plagiarism Checker

The submitted paper should report original, previously unpublished research results of the experiment or theory. The paper submitted to the publishers should meet their criteria and must not be under consideration for publication elsewhere. Since ethical conduct is the most essential virtue of any academic, any act of plagiarism is unacceptable. If you are found to commit plagiarism, the following acts of sanction would be taken:

- The publisher will reject your paper or delete the article from the final publication.
- The publisher will report the violation to you and your affiliation.
- The publisher will inform the violation to your research funding agency.

There are many plagiarism-checking programs on the Internet:

- Turnitin (www.turnitin.com)
- Article Checker (www.articlechecker.com)

- Scan My Essay (www.scanmyessay.com)
- Plagiarism Detect (http://plagiarism-detect.com)
- Plagiarism (www.plagiarism.org)
- Grammarly (www.grammarly.com)
- Smallseotools (http://smallseotools.com/plagiarism-checker)

Although most of them are free, the quality of checks is not high. Here I will introduce the plagiarism checker Turnitin, which requires an annual fee. Note that a paper is accepted if the similarity of content is less than 25–30%.

The instructions for using the TURNITIN program are as follows:

1. Click at http://www.turnitin.com.
2. Create a new account in case you are a new user.
3. At "Create a User Profile," click the "Instructor" icon under "Create a New Account."
4. At "Create a New Instructor Account," key in the Account ID of your university, password, your name, e-mail address, secret question and answer. Then, click "I Agree – Create Profile."
5. Click "Add Class."
6. At "Create a New Class," click "Standard," write class name, and click on "Submit."
7. Click the "Class Name" to enter the class and start creating assignments by clicking on "Add Assignment."
8. At "New Assignment," write the assignment title. After that click on "Optical Setting."
9. Click on "No Repository" and then on "Submit."
10. Chick on "Add Assignment," "View," and "Submit File."
11. Upload the file and then click on "Upload" and "Confirm."
12. Click on the "Go to Assignment" box.
13. Click on the "View" link of your desired assignment. Here you will see percentage similarity of the report.

7.2.1 Example of Research Paper Compared by Plagiarism Checker

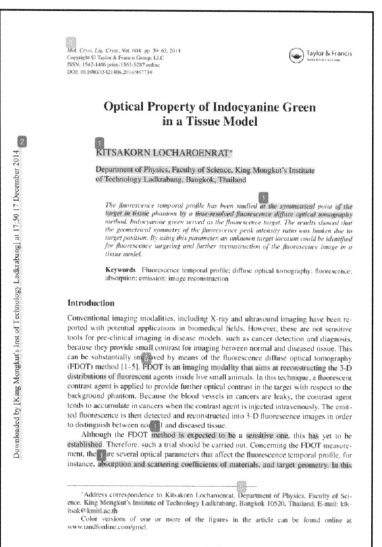

60/[316]   *K. Locharoenrat*

paper, the focus is on fluorescence temporal profile changes, relying on target position (centered to the near boundary). A time-correlated single photon counter served as a detection system. First, fluorescence temporal profiles are presented as a function of geometrical configuration at the symmetrical point of indocyanine green target in tissue phantom. Next, an unknown target position was identified using the fluorescence peak intensity ratio. This analysis should be useful to decide the excitation-detection geometry for further aims in optical reconstruction imaging in a tissue model.

Experimental

A fluorescence target of 6 mm in diameter was performed with different target positions (4.5, 7.5, or 15 mm, apart from phantom surface) and concentrations (0.5, 1.0, and 2.0 mM indocyanine green dye in 1% intralipid solution). This target was filled in a tissue phantom in which the top view is shown in Fig. 1. The target was then excited by incident light at position A (90^0). The light source was a Ti:Sapphire laser with a central wavelength of 780 nm. Fluorescence light was detected by a photomultiplier. Detection points were focused at the symmetrical point B (30^0) and D (-30^0), as shown in Fig. 1.

Results and Discussion

Figures 2 (a), (b), and (c) show the fluorescence temporal profiles from a 1.0 mM Indocyanine green (ICG) target 6 mm in diameter with different target locations, namely d = 4.5, 7.5, and 15 mm, apart from the phantom surface. The difference of the fluorescence temporal profile at symmetrical point B (30^0) and D (-30^0) became large when the target was near the phantom boundary or point C (0^0). This was because the total path length

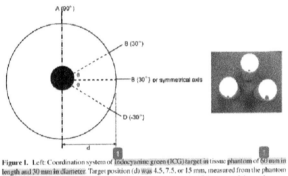
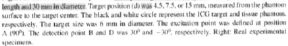

Figure 1. Left: Coordination system of indocyanine green (ICG) target in tissue phantom of 60 mm in length and 30 mm in diameter. Target position (d) was 4.5, 7.5, or 15 mm, measured from the phantom surface to the target center. The black and white circle represent the ICG target and tissue phantom, respectively. The target size was 6 mm in diameter. The excitation point was defined at position A (90^0). The detection point B and D was 30^0 and -30^0, respectively. Right: Real experimental specimens.

Optical Property of ICG in a Tissue Model [317]/6]

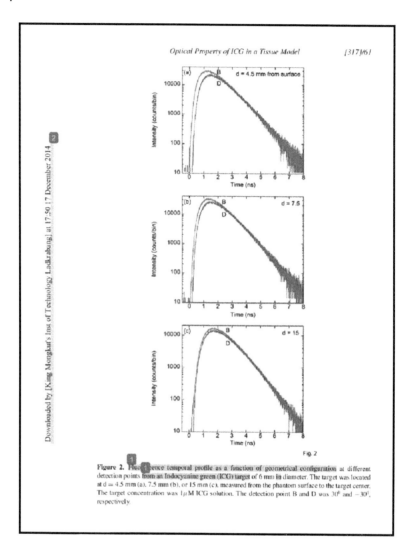

Fig. 2

Figure 2. Fluorescence temporal profile as a function of geometrical configuration at different detection points from an Indocyanine green (ICG) target of 6 mm in diameter. The target was located at d = 4.5 mm (a), 7.5 mm (b), or 15 mm (c), measured from the phantom surface to the target center. The target concentration was 1 μM ICG solution. The detection point B and D was 30° and −30°, respectively.

62/[318]   *K. Locharoenrat*

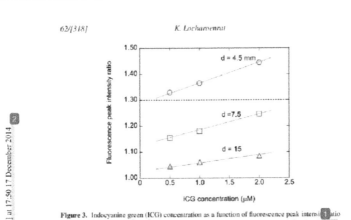

ICG concentration (µM)

Figure 3. Indocyanine green (ICG) concentration as a function of fluorescence peak intensity ratio at a detection point of $30^0/-30^0$. The target was 6 mm in diameter and was located at d = 4.5 mm, 7.5 mm, or 15 mm, measured from the phantom surface to the target center.

variation between excitation light and emission light in the tissue phantom was increased. Furthermore, the peak intensity and rising edge of fluorescence temporal profile at 30^0 were higher and earlier than those of the profile at -30^0. This result suggested that the front region of the target was more excited than the backside region. Then, the emitted fluorescence travelled a short path length to the detection point. On the other hand, typically the fluorescence intensity would decay as $I(t) = I_0 \exp(-t/t)$, where t is a fluorescent lifetime. By fitting the observed fluorescence-decay profiles to: $I(t) = I_0 \exp(-t/t)$, one can obtain t = 0.5 ns.

Next, we characterized the fluorescence temporal profile in terms of fluorescence peak intensity ratio, as shown in Fig. 3. When the target came close to the phantom boundary, the degree of asymmetry in terms of fluorescence peak intensity ratio was further from unity. This indicated that the symmetrical plane of the target with respect to point A (90^0) was broken. Moreover, the degree of this asymmetry was increased when the ICG concentration was increased. Increasing the absorption coefficient of ICG enhanced a gradient of emission intensity at the target region.

These basic results showed a promising approach for fluorescence targeting and further reconstruction of the fluorescence image. The asymmetry of the fluorescence peak intensity ratio was systematically dependent on the target depth and concentration. Therefore, a complete set using this parameter was able to identify an unknown target location. For example, the target size and concentration were known to be 6 mm and 0.5 – 2.0 mM ICG solution, respectively. If the fluorescence peak intensity ratio was beyond 1.3, an unknown target position was expected to be about 4.5 mm or more from the phantom boundary. By contrast, if the fluorescence peak intensity ratio was below 1.3 (dash line), an unknown target point tended to be deeper than 7.5 mm, apart from the phantom surface.

This asymmetrical feature suggests that the fluorescence peak intensity ratio carries geometrical information about the target location. The image quality of the target could

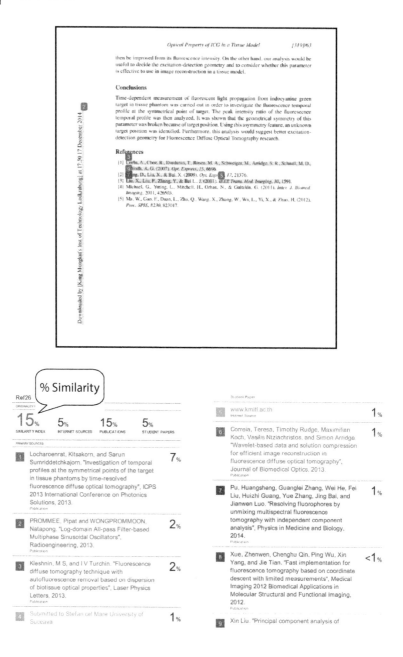

% Similarity

Ref26

ORIGINALITY

15%	5%	15%	5%
SIMILARITY INDEX	INTERNET SOURCES	PUBLICATIONS	STUDENT PAPERS

PRIMARY SOURCES

1 Locharoenrat, Kitsakorn, and Sarun Sumriddetchkajorn. "Investigation of temporal profiles at the symmetrical points of the target in tissue phantoms by time-resolved fluorescence diffuse optical tomography", ICPS 2013 International Conference on Photonics Solutions, 2013. *Publication* — 7%

2 PROMMEE, Pipat and WONGPROMMOON, Natapong. "Log-domain All-pass Filter-based Multiphase Sinusoidal Oscillators", Radioengineering, 2013. *Publication* — 2%

3 Kleshnin, M S, and I V Turchin. "Fluorescence diffuse tomography technique with autofluorescence removal based on dispersion of biotissue optical properties", Laser Physics Letters, 2013. *Publication* — 2%

4 Submitted to Stefan cel Mare University of Suceava — 1%

Student Paper

5 www.kmitl.ac.th *Internet Source* — 1%

6 Correia, Teresa, Timothy Rudge, Maximilian Koch, Vasilis Ntziachristos, and Simon Arridge. "Wavelet-based data and solution compression for efficient image reconstruction in fluorescence diffuse optical tomography", Journal of Biomedical Optics, 2013. *Publication* — 1%

7 Pu, Huangsheng, Guanglei Zhang, Wei He, Fei Liu, Huizhi Guang, Yue Zhang, Jing Bai, and Jianwen Luo. "Resolving fluorophores by unmixing multispectral fluorescence tomography with independent component analysis", Physics in Medicine and Biology, 2014. — 1%

8 Xue, Zhenwen, Chenghu Qin, Ping Wu, Xin Yang, and Jie Tian. "Fast implementation for fluorescence tomography based on coordinate descent with limited measurements", Medical Imaging 2012 Biomedical Applications in Molecular Structural and Functional Imaging, 2012. *Publication* — <1%

9 Xin Liu. "Principal component analysis of

7.3   Problems

Pb.1. Show the difference between research ethics concerning participants and researchers.

Pb.2. Check the originality of a paper of your interest by using the Plagiarism and Turnitin programs. Discuss the difference between the methods adopted by these two programs.

Pb.3. Write your own research paper based on the following template and check the originality of your paper.

Research conceptualization process

1	Title *Write a few tentative titles and select the best one later.*
2	Abstract *Write an abstract when the whole document is completed.*
3	Keywords
4.1	State a conceptual framework *Show a block diagram and connect your variables to the theories.*
4.2	Literature review based on the research problem to solve/explain/test and develop. *Provide evidence of the existence of the problem.* *Cite the previous studies that are inconclusive or contradictory.* *Prove that there is a lack of studies in this area and that it is important to fill this gap.*
4.3	Objective *Based on the research problem, state what your study intends to do and what is the significance of this study.*
5	Methodology *What are the samples, sampling, and instruments?*
6.1	Results *State the results based on each research problem.*
6.2	Discussion *Compare your findings with the findings from previous studies reviewed in the literature review. Give reasons for why your findings support/do not support the previous findings.*
7	Conclusion
8	References

Chapter 8

Presentation Format

After your paper is accepted, the presentation of your work (although optional) can help you share the idea and/or get financial support from audiences/readers. In a conference, seminar, or workshop, you can select your preferred presentation format during the online abstract submission process: oral or poster presentation. On the other hand, you can present your research proposal for receiving financial support when you would like to carry out further studies in the future.

8.1 Oral Presentation

For oral presentation, the time generally allocated for a plenary speaker is 45 min, for a keynote speaker is 25 min, and for a contributory speaker is 15 min, which accommodates the presentation followed by a short discussion. You should prepare your presentation with 1–2 slides/min. The content of the presentation is similar to manuscript submission: Introduction, Experiment, Results and Discussion, and Conclusion.

The following guidelines are very useful for oral presentation when you present your work in a conference, seminar, or workshop:

1. Introduce yourself, your advisor, and committee members.

Research Methodologies for Beginners
Kitsakorn Locharoenrat
Copyright © 2017 Pan Stanford Publishing Pte. Ltd.
ISBN 978-981-4745-39-0 (Hardcover), 978-1-315-36456-8 (eBook)
www.panstanford.com

2. Give an introduction and background information on your topic. What relevant research has been performed previously?
3. State the problems that remain unanswered.
4. State your objectives clearly and give the specific hypotheses you wish to show.
5. Describe the methodology. Be sure you fully understand your chosen methods. Give reasons why you chose these methods over other approaches.
6. Present any data you have collected.
7. Explain the significance of your findings or potential future findings.

8.1.1 Examples of Oral Presentation

Oral Presentation: Example 1

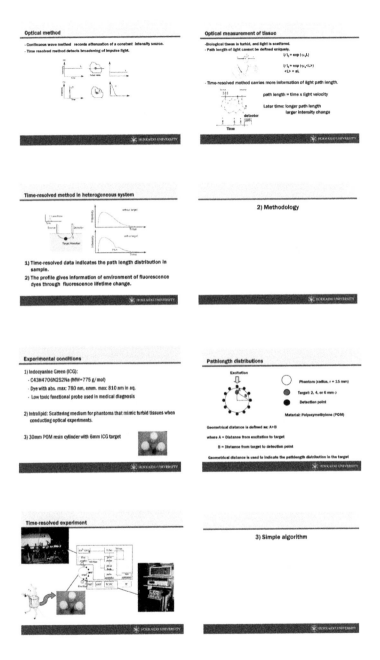

Optical method

- Continuous wave method records attenuation of a constant intensity source.
- Time resolved method detects broadening of impulse light.

Optical measurement of tissue

- Biological tissue is turbid, and light is scattered.
- Path length of light cannot be defined uniquely.

$I/I_0 = \exp(-\mu_a L)$

$I/I_0 = \exp(-\mu_a <L>)$

$<L> = \alpha L$

- Time-resolved method carries more information of light path length.

path length = time x light velocity

Later time: longer path length
larger intensity change

detector

Time

Time-resolved method in heterogeneous system

1) Time-resolved data indicates the path length distribution in sample.
2) The profile gives information of environment of fluorescence dyes through fluorescence lifetime change.

2) Methodology

Experimental conditions

1) Indocyanine Green (ICG):
   - C43H4706N2S2Na (MW=775 g/mol)
   - Dye with abs. max: 780 nm, emm. max: 810 nm in aq.
   - Low toxic functional probe used in medical diagnosis

2) Intralipid: Scattering medium for phantoms that mimic turbid tissues when conducting optical experiments.

3) 30mm POM resin cylinder with 6mm ICG target

Pathlength distributions

Excitation

Phantom (radius, r = 15 mm)

Target: 2, 4, or 6 mm ⌀

Detection point

Material: Polyoxymethylene (POM)

Geometrical distance is defined as: A+B

where A = Distance from excitation to target

B = Distance from target to detection point

Geometrical distance is used to indicate the pathlength distribution in the target

Time-resolved experiment

3) Simple algorithm

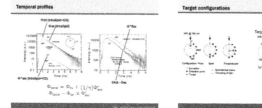

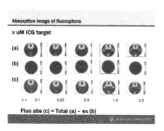

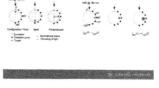

Oral Presentation: Example 2

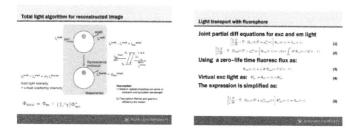

Optical Delay Line for Rapid Scanning Low-Coherence Reflectometer

Kitsakorn Locharoenrat[1]* and I-Jen Hsu[2]

[1]Department of Physics, Faculty of Science
King Mongkut's Institute of Technology Ladkrabang, Thailand
[2]Department of Physics, Faculty of Science
Chung Yuan Christian University, Taiwan ROC

CONTENTS

I. INTRODUCTION - BACKGROUND

Tomography: Imaging interior structure of a substance w/o invasion
- Identifying locations and profiles of constituents.

Optical Coherence Tomography (OCT) system based on correlation techniques comparing reflected light signal of sp-ref beam.

I. INTRODUCTION - BACKGROUND

× Different designs of delay lines proposed
- Rotating cube [ballif et al Opt Lett 22 (1997) 757]

- Polygonal scanner [Delachenal et al Opt Comm 162 (1999) 195]

But, these configurations are difficult to fabricate in small size!

I. INTRODUCTION - PURPOSES

To construct optical delay line which is compact and easy to fabricate.

To improve scanning rate.

To improve scanning range.

II. MATERIALS AND METHODOLOGIES

Fabrication of optical delay line
1. Retro-reflector (2x2cm)

2. Scanning mirror (400 Hz scanning rate, 0-10° tilted angle)

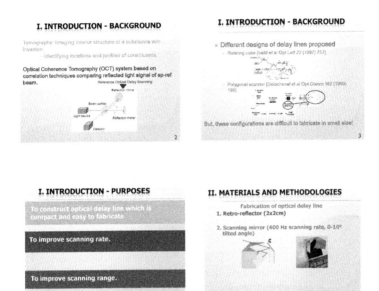

II. Materials and methodologies

Optical setup

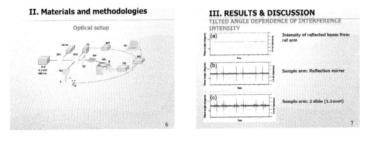

III. RESULTS & DISCUSSION

TILTED ANGLE DEPENDENCE OF INTERFERENCE INTENSITY

(a) Intensity of reflected beam from ref arm

(b) Sample arm: Reflection mirror

(c) Sample arm: 2 slide (1.1mm)

IV. CONCLUSION

1. We constructed the optical delay line for rapid scanning.

2. Scanning rate was improved at 400Hz.

3. Scanning range was improved more than 3mm.

Presentation: Example 3

Second-Order Nonlinear Optical Response of
Metal Nanostructures

Kitsakorn Locharoenrat

Department of Physics, Faculty of Science
King Mongkut's Institute of Technology Ladkrabang
Bangkok 10520 Thailand

Contents

I. **Introduction:** Background, Purposes

II. **Literature review:** Nanofabrication, Optical properties

III. **Materials & methodologies:** Sample preparation, Nonlinear optical experimental setup

IV. **Results & discussion**

V. **Conclusion**

1

Nanotechnology: undergoing a period of explosive growth

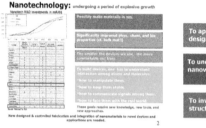

New designed & controlled fabrication and integration of nanomaterials to novel devices and applications are needed.

2

I. Introduction - Motivation

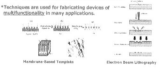

To apply scientific knowledge of building designed nanostructure in bottom-up process.

To understand the self organization of metallic nanowires.

To investigate properties of plasmon resonant structures in metallic nanowires.

3

I. Introduction - Purposes

* Fabricate metallic nanowires by a shadow deposition method.

* Observe SH signal from metallic nanowires by using SH spectroscopy.

* Compare optical properties from metallic nanowires.

4

II. Literature review - Nanofabrication

*There are numerous methods to synthesize nanomaterials of various characteristics.

* An essential challenge in synthesis is to control structures at a high yield for industrial applications.

* Techniques are used for fabricating devices of multifunctionality in many applications.

Membrane-Based Template   Electron Beam Lithography

5

II. Literature review – Shadow deposition

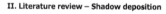

Merits: material diversity, high throughput, and potential for high-vol. production.

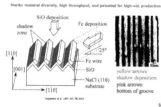

SiO deposition

shadow zone   Fe deposition

[110]
[001]
Fe wire
SiO
[110]   NaCl (110) substrate

yellow arrows: shadow deposition
pink arrows: bottom of groove

6

II. Literature review - Optical properties

Second Harmonic Generation (SHG)

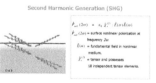

$\vec{P}_{sur}(2\omega) = \varepsilon_0 \hat{\chi}^{(2)} : \vec{E}(\omega)\vec{E}(\omega)$

$\vec{P}_{sur}(2\omega)$ = surface nonlinear polarization at frequency 2ω.

$\vec{E}(\omega)$ = fundamental field in nonlinear medium.

$\hat{\chi}^{(2)}$ = tensor and possesses 18 independent tensor elements.

7

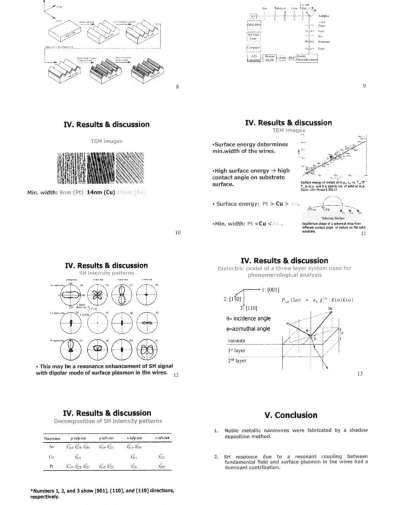

8.2 Poster Presentation

Poster presentation is a casual demonstration of a novel and applicable idea in a simple and concise manner. It is an effective mode for discussion and for receiving responses, which can help you to refine your research and develop new ideas for future studies. A poster is generally displayed during a conference. It should fit within the offered area. Each poster is normally allocated a board of width 841 mm and height 1189 mm. The presenter should stand beside the poster for about 1–2 h during the display period. A poster has the following format:

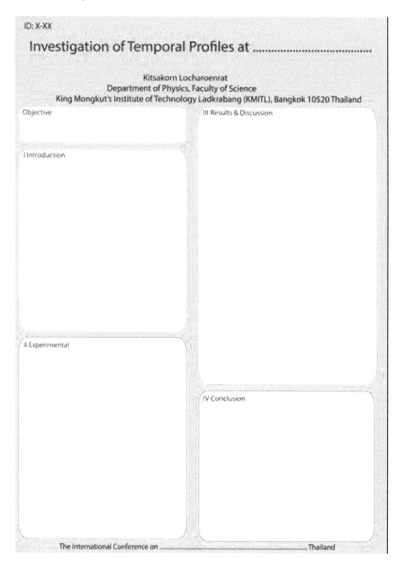

8.2.1 Examples of Poster Presentation

ID: 16

Field Emission from Tungsten Nanotips

Kitsakorn Locharoenrat

Department of Physics, Faculty of Science

King Mongkut's Institute of Technology Ladkrabang (KMITL), Bangkok 10520 Thailand

Objective

To study field emission characteristics of tungsten tips, in addition to a metal grating [1], for generation of terahertz radiation.

I Introduction

Fowler-Nordheim (F-N) theory with an enhanced value of electric field;

$$J \propto 1.5 \times 10^{-6} \, (\beta E)^2 / \emptyset \, \exp(-6.8 \times 10^9 \, \emptyset^{1.5} / \beta E)$$

Fig. 1 Potential energy diagram for field emission (FE) from metal.

Causes of laser-induced FE
* Absorbed photon
* Laser-induced FE
* Heating

Fig. 2 Field emission with I-E characteristics.

Metal nanostructures incl. tungsten tip can increase the enhancement factor, leading to lower turn-on voltage for FE --> by controlling the tip radius.

II Experimental

@ Tungsten tip was formed by electrochemical etching in 1 M NaOH solution.

Fig. 3 Etching mechanism around air-sol interface.

@The tips were observed by SEM.

@ FE characteristics were measured in UHV chamber evacuated to 10^{-8} Torr.

Fig. 4 Field emission set-up.

II Experimental (Cont d)

Fig. 5 Field emission set-up inside the UHV chamber.

III Results & Discussion

Tungsten nanotips with r = 10-50 nm were produced.

Turn-on electric field at an emission current of 1pA is at a potential of E ~ 5 V/μm.

Field enhancement factor β of the tip ~ 500.

Field emission was observed at field F ~ 2 V/nm.

Fig 6. FE measurements of tungsten tip; (a) Current density (J) vs. applied potential (E) and (b) corresponding F-N plot.

IV Conclusion

+ Tungsten tips were produced by electrochemical etching with 10-50 nm in diameter range.

+ Field emission test on the fabricated tungsten tips were observed.

Future Work

- Fs laser-induced field emission.

- Terahertz radiation from the fabricated tungsten tips.

Fig 7 Terahertz pulse emitted from tungsten tip due to ultrafast pulses at 800nm.

[1] Gregor M Welsh, Neil T Hunt, and Klaas Wynne, PRL 98 (2007), 026803

The 8th Asian Meeting on Ferroelectrics, Dec 9-14, 2012, Pattaya, Thailand

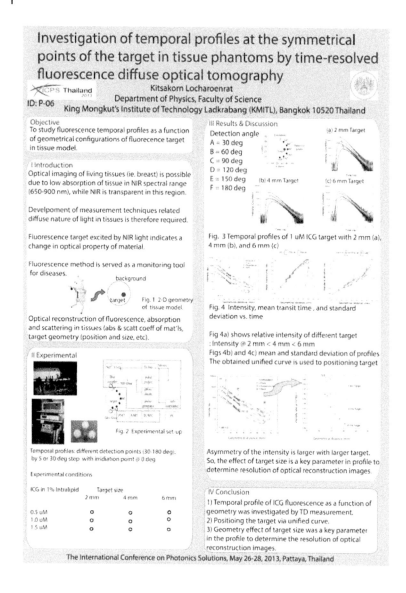

Investigation of temporal profiles at the symmetrical points of the target in tissue phantoms by time-resolved fluorescence diffuse optical tomography

ICPS Thailand

ID: P-06

Kitsakorn Locharoenrat
Department of Physics, Faculty of Science
King Mongkut's Institute of Technology Ladkrabang (KMITL), Bangkok 10520 Thailand

Objective
To study fluorescence temporal profiles as a function of geometrical configurations of fluorecence target in tissue model.

I Introduction
Optical imaging of living tissues (ie. breast) is possible due to low absorption of tissue in NIR spectral range (650-900 nm), while NIR is transparent in this region.

Develpoment of measurement techniques related diffuse nature of light in tissues is therefore required.

Fluorescence target excited by NIR light indicates a change in optical property of material.

Fluorescence method is served as a monitoring tool for diseases.

Fig. 1 2-D geometry of tissue model.

Optical reconstruction of fluorescence, absorption and scattering in tissues (abs & scatt coeff of mat'ls, target geometry (position and size, etc).

II Experimental

Fig. 2 Experimental set up

Temporal profiles: different detection points (30-180 deg), by 5 or 30 deg step with irridiation point @ 0 deg

Experimental conditions

ICG in 1% Intralipid	Target size		
	2 mm	4 mm	6 mm
0.5 uM	o	o	o
1.0 uM	o	o	o
1.5 uM	o	o	o

III Results & Discussion

Detection angle
A = 30 deg
B = 60 deg
C = 90 deg
D = 120 deg
E = 150 deg
F = 180 deg

(a) 2 mm Target
(b) 4 mm Target
(c) 6 mm Target

Fig. 3 Temporal profiles of 1 uM ICG target with 2 mm (a), 4 mm (b), and 6 mm (c)

Fig. 4 Intensity, mean transit time , and standard deviation vs. time

Fig 4a) shows relative intensity of different target : Intensity @ 2 mm < 4 mm < 6 mm
Figs 4b) and 4c) mean and standard deviation of profiles The obtained unified curve is used to positioning target

Asymmetry of the intensity is larger with larger target. So, the effect of target size is a key parameter in profile to determine resolution of optical reconstruction images.

IV Conclusion
1) Temporal profile of ICG fluorescence as a function of geometry was investigated by TD measurement.
2) Positioing the target via unified curve.
3) Geometry effect of target size was a key parameter in the profile to determine the resolution of optical reconstruction images.

ID:

Optical Investigation of Palladium-Coated Gold Nanorods

Kitsakorn Locharoenrat and Pattareeya Kittidachachan

Department of Physics, Faculty of Science

King Mongkut's Institute of Technology Ladkrabang (KMITL), Bangkok 10520 Thailand

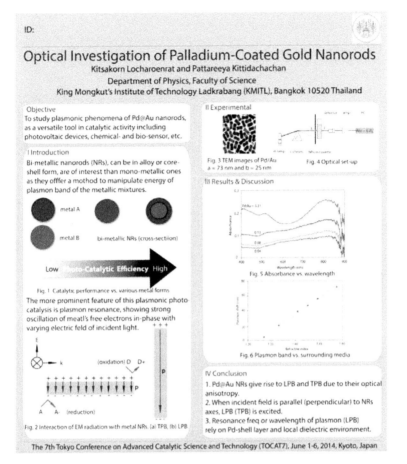

Objective

To study plasmonic phenomena of Pd@Au nanorods, as a versatile tool in catalytic activity including photovoltaic devices, chemical- and bio-sensor, etc.

I Introduction

Bi-metallic nanorods (NRs), can be in alloy or core-shell form, are of interest than mono-metallic ones as they offfer a method to manipulate energy of plasmon band of the metallic mixtures.

metal A

metal B   bi-metallic NRs (cross-section)

Low Photo-Catalytic Efficiency High

Fig. 1 Catalytic performance vs. various metal forms

The more prominent feature of this plasmonic photo-catalysis is plasmon resonance, showing strong oscillation of meatl's free electrons in-phase with varying electric feld of incident light.

(oxidation) D   D+

A   A-   (reduction)

Fig. 2 Interaction of EM radiation with metal NRs. (a) TPB, (b) LPB.

II Experimental

Fig. 3 TEM images of Pd/Au   Fig. 4 Optical set-up
a = 73 nm and b = 25 nm

III Results & Discussion

Fig. 5 Absorbance vs. wavelength

Fig. 6 Plasmon band vs. surrounding media

IV Conclusion

1. Pd@Au NRs give rise to LPB and TPB due to their optical anisotropy.
2. When incident field is parallel (perpendicular) to NRs axes, LPB (TPB) is excited.
3. Resonance freq or wavelength of plasmon (LPB) rely on Pd-shell layer and local dielectric environment.

The 7th Tokyo Conference on Advanced Catalytic Science and Technology (TOCAT7), June 1-6, 2014, Kyoto, Japan

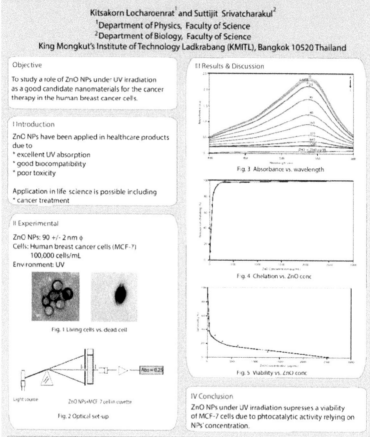

ID: T-13

Ultraviolet Irradiation Induced Enhancement in Functional Zinc Oxide Nanostructure Exposure against Haman Cancer Cells

Kitsakorn Locharoenrat[1] and Suttijit Srivatcharakul[2]

[1] Department of Physics, Faculty of Science
[2] Department of Biology, Faculty of Science
King Mongkut's Institute of Technology Ladkrabang (KMITL), Bangkok 10520 Thailand

Objective

To study a role of ZnO NPs under UV irradiation as a good candidate nanomaterials for the cancer therapy in the human breast cancer cells.

I Introduction

ZnO NPs have been applied in healthcare products due to
* excellent UV absorption
* good biocompatibility
* poor toxicity

Application in life science is possible including
* cancer treatment

II Experimental

ZnO NPs: 90 +/- 2 nm φ
Cells: Human breast cancer cells (MCF-7)
100,000 cells/mL
Environment: UV

Fig. 1 Living cells vs. dead cell

Light source   ZnO NPs+MCF-7 cell in cuvette

Fig. 2 Optical set-up

III Results & Discussion

Fig. 3 Absorbance vs. wavelength

Fig. 4 Chelation vs. ZnO conc

Fig. 5 Viability vs. ZnO conc

IV Conclusion

ZnO NPs under UV irradiation supresses a viability of MCF-7 cells due to phtocatalytic activity relying on NPs' concentration.

The 16th International Conference on Thin Films (ICTF-16), October 13-16, 2014, Dubrovnik, The Republic of Croatia

ID:

Preparation and Characterization of Zinc Oxide Nanoparticles

Kitsakorn Locharoenrat
Department of Physics, Faculty of Science
King Mongkut's Institute of Technology Ladkrabang (KMITL), Bangkok 10520 Thailand

Objective

To prepare of ZnO NPs using pulse laser ablation technique served as photocatalyst.

I Introduction

Traditional methods to produce NPs have some drawbacks
* technical simplicity
* low yield
* contamination

Novel method to prepare NPs using pulsed-laser is possible.

II Experimental

Target: Zn plate
Environment: Sodium Dodecyl Sulfate
Light source: Nd:YAG laser (1064 nm, 100 mJ)
Spectometer: 300 - 600 nm

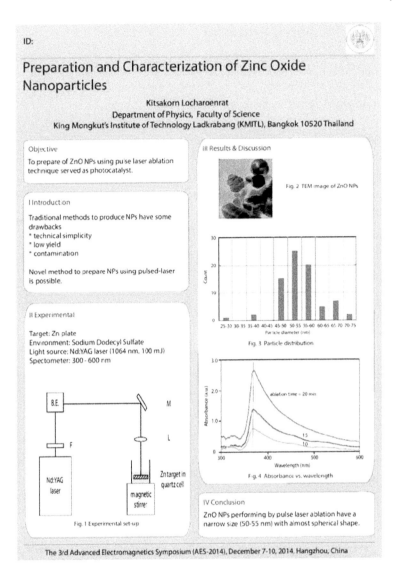

Fig. 1 Experimental set-up

III Results & Discussion

Fig. 2 TEM image of ZnO NPs

Fig. 3 Particle distribution

Fig. 4 Absorbance vs. wavelength

IV Conclusion

ZnO NPs performing by pulse laser ablation have a narrow size (50-55 nm) with almost spherical shape.

The 3rd Advanced Electromagnetics Symposium (AES-2014), December 7-10, 2014, Hangzhou, China

8.3   Research Proposal

A research proposal is a request for supporting the sponsored research, instruction, or extension projects. The proposal for sponsored activities generally follows a format similar to that of the manuscript, although there are variations depending on whether the proposer is seeking support for a research grant, a conference, or curriculum development project. The following outline covers the primary components of a research proposal: Title, Aim, Background and Outline, Method, Schedule, and References. Your proposal will be a variation on this basic outline.

8.3.1   Examples of Research Proposal

Research Proposal: Example 1

Title: Synthesis and Characterization of Novel Nanomaterials for Biosensor Applications

Aim

This project is to synthesize nanowires and nanoclusters, and to characterize them using analytical techniques, such as Raman, IR, and transmission electron microscopy, as well as to apply them for biosensor applications.

Background

We are going to employ biomaterials such as amyloid fibrils served as templates for a fabrication in a bottom-up approach to synthesize the novel nanowires. The nanomaterials characterization would be performed by using Raman, IR, and transmission electron microscopy (TEM) technique. Finally we would measure the physical properties of these synthesized nanowires by measuring zeta potentials, conductance, and thermal properties. As a continuation of the project to synthesize nanomaterials, we would also synthesize nanoclusters of different shapes for biosensor applications. We would closely study the growth of these clusters and a unique property by using analytical techniques, such as UV, IR, Raman, and TEM.

Project Outline

Fabrication of nanomaterials is one of the most challenging researches in a field of materials science. Biological materials such as proteins have a unique property of naturally adhering to the inorganic ions in nature. By utilizing this unique property, we could obtain the amyloid fibrils from denaturating the proteins by heating it over an elevated temperature. The silver and copper nanoparticles are therefore attached over a surface of these fibrils by a special technique so-called electroless plating. Alternatively a conjugation of copper or silver nanoparticles could be carried out in a solution phase. Thus, we could obtain the ultrathin copper and silver nanowires of different configurations and sizes having a unique property of high conductivity, flexibility, and thermal stability. These could be employed for making lab-on-chip devices which can be used for biosensor applications. Next, the nanoclusters would be produced by depositing the platinum or palladium nanoparticles using electrochemical process over the pre-formed silica support at an elevated temperature for extended hours by employing suitable surfactants, such as CTAB or sodium citrate, as directors for particular form of growth for (111) plane of growth or (100) plane of growth. We could vary a ratio of reducing agent and surfactants to control the growth of the nanoclusters. Finally, the characterization of these synthesized nanomaterials (nanowires and nanoclusters) could be carried out by using TEM techniques, and also by using many spectroscopy techniques (i.e., UV, IR, Raman techniques).

Raw Materials

1. $CuSO_4 \cdot 5H_2O$
2. $AgNO_3$
3. Sodium borohydride (reducing agent)
4. Sodium citrate (stabilizer)
5. Nitric acid
6. Ultrapure water
7. Copper and nickel grid supports
8. Proteins (lysozyme, insulin, or BSA)
9. Thermostat
10. pH meter
11. Magnetic stirrer
12. Centrifugator
13. Analytical instruments (UV, IR, TEM)

Methods

1. *Synthesis of nanowires*: The synthesis of nanowires could be carried out in two methods. The first one is the electroless plating. In this method, we deposit the amyloid fibrils over the top of carbon-coated nickel grid support, and then reducing agents (i.e., sodium borohydride) and stabilizers (i.e., CTAB or sodium citrate) are added. Hence, copper or silver nanoparticles are synthesized over the top of these deposited amyloid fibrils. Another method is to allow the conjugation of the nanoparticles on the top of the amyloid fibrils by allowing it to remain in an eppendorf tube for overnight at room temperature and then centrifuging and removing unreacted nanoparticles. By this method, we are able to synthesize nanowires ranging from 400 nm to several micrometers of different configurations and these nanowires could be characterized by using TEM. The same biotemplated nanowires could be characterized by using UV, IR, and Raman spectroscopy technique. By using Raman, we could also probe the interaction of individual nanoparticles over the surface of the amyloid fibrils. These information are very valuable to study the interaction between inorganic and biomolecules in forming the nanowires.

2. *Synthesis of nanoclusters*: The nanoclusters could be synthesized by depositing the individual nanoparticles (i.e., platinum, palladium, or silver nanoparticles) over the surface of the alumina or silica support. This could be performed by electrochemical deposition technique. The growth of the nanoclusters could be studied by using different techniques, such as transmission electron microscope, UV, IR, and Raman spectroscopy techniques. This is because each nanocluster would have a definite shape and a route of conjugation of the nanoparticles to form nanoclusters. Then, the orientation of the lattice planes could be understood to probe the structural morphology of these nanoclusters. Additionally, AFM could be used to study the topology of these synthesized nanoclusters and elemental analysis of these nanoclusters could be performed by using TEM techniques.

Schedule

The project period is 24 months and the total time of this project could be visualized as follows:

Literature review	10 weeks
Experiment	30 weeks

Data analysis	20 weeks
Reproducibility	16 weeks
Physical property measurements	20 weeks

Research Proposal: Example 2

Title: Magnetic Anisotropy of the Co Nanowires on the GaAs Substrates

Aim

Our purpose is to study the Co nanostructures on the semiconductor substrates which are expected to exhibit the magnetic characterization on the Co nanowires. The Co thickness dependence and the growth temperature dependence on the magnetic properties of the Co nanowires would be investigated through in situ Brillouin light scattering spectroscopy.

Background

The study of the magnetic metallic nanostructures has contributed a wide range of new phenomena. The 3-d metals are considered as the ferromagnetism in which the magnitude of the magnetic moment does rely on the electron density. Additionally the electron number changes the Fermi level variation of the magnetization. These have drawn attention of the scientists involved in the research of the basic aspects of low dimensional systems and the materials scientists who intend to apply magnetic nanostructures into devices. The hybrid ferromagnet–semiconductor structures offer several unique benefits to the magnetoelectronic applications. Magnetoelectronics is a new field that is devoted to the invention and development of electrical device structures that incorporate a ferromagnetic element. Generally, the magnetic anisotropy is a fundamental property in determining the magnetic behavior of thin and ultrathin film magnets. From a theoretical point of view, an anisotropy plays an important role in stabilizing ferromagnetism in 2-d magnets. The magnetic anisotropies have been observed in a wide range of ultrathin magnetic films made of the 3-d metals. Mewes et al. [1] have studied the magnetic anisotropies of the Co films on MgO substrates and revealed that the resulting magnetic behavior is the same high level as the Co films grown on the Cu substrates proposed by Fassbender et al. [2]. Frank et al. have shown the uniaxial magnetic anisotropy contribution to the Fe films on the Ag(001)/GaAs(001) substrates [3]. Rickart et al. have studied the magnetic anisotropies of the Fe films on the

Ag(001)/MgO(001) substrates [4]. These results have led us to extend investigation of Co nanowires on the semiconductor substrate because it has the possibility of dilute magnetic semiconductors (DMSs) as they combine the two fields in the condensed matter physics: magnetism and semiconductor. DMSs are the key material of the spintronics that applies the correlation between charge and spin of electrons to devices. The surface effects in the metallic nanoparticles are of great importance because the magnetic properties of the nanoparticles are strongly influenced when their surface is in contact with different media. Consequently, the proof that Co nanowires on the GaAs substrates are DMSs would give a large contribution to the spintronics because the Curie temperature of the existing ferromagnetism DMSs, such as the Mn films on the GaAs substrates, is too low to be used for the spintronics devices. To elucidate the magnetic origin of the Co nanowires on the GaAs substrates, the Brillouin light scattering spectroscopy would be introduced for this study in order to observe a magnetic anisotropy in the Co nanowires deposited onto the GaAs substrates. Although the light scattering in solid can be probed with many techniques, such as Raman scattering, referring to the scattering from high frequency excitations, the experimental techniques used for Brillouin scattering are more sophisticated than those for Raman scattering partly due to the need to be able to detect much smaller frequency shifts. Single-mode lasers must be used to ensure that the laser linewidth is sufficiently small, and scanning Fabry–Perot interferometer is used instead of a grating spectrometer to obtain the required frequency resolution. Therefore, we use in situ Brillouin light scattering spectroscopy to study the magnetic properties of the Co nanowires. The relation between magnetism and the Co nanoparticle characteristics (size-dependent) will be also investigated.

Project Outline

The Co nanowires deposited onto the GaAs substrates would be shown a transition of the magnetic anisotropy depending on its thickness. The nanostructures are deposited onto the GaAs substrates by the EBL machine in the UHV conditions in a standard MBE system. The thickness and the temperature dependence of magnetism would be then investigated by in situ Brillouin light scattering spectroscopy.

Methods

Using microfabrication techniques developed in the field of semiconductor physics, nanoscale V-shaped groove structures can be constructed on the surface of the GaAs wafer; 100–200 nm pitch with 10 mm × 10 mm patterned area of V-groove shape. The (100) oriented GaAs wafer is masked by photoresist and then a stripe pattern, and is printed by using electron beam lithography. By wet etching rate, V-shaped grooves are formed. After such a microstructured substrate is available, the Co nanowires will be prepared by an electron beam lithography (EBL) machine in the ultrahigh vacuum (UHV) on the GaAs layer in the multichamber molecular beam epitaxy (MBE) system. The Co nanowires are deposited onto this substrate, and the temperature is kept constant during the deposition. The process of the growth is confirmed by taking the in situ reflection high energy electron diffraction (RHEED) images. The nanowires thickness ranging from 10 nm to 100 nm at each growth temperature are measured by a quartz crystal oscillator, which is calibrated by an atomic force microscope (AFM). It is well known that changes in substrate preparation and growth conditions could cause significant changes in the magnetic properties of the nanowires. In order to carry out the thickness dependent magnetic measurements without having such impact, the nanowires are prepared exactly on the same substrate under an identical growth conditions. Then the magnetism of the nanowires is checked by in situ Brillouin light scattering spectroscopy.

For the magnetic measurements, we use in situ Brillouin light scattering spectroscopy. The laser light is passed through a beam expander, a polarizer, a beam splitter before being focused in the sample by lens. Finally the scattered light is analyzed by a Fabry–Perot interferometer.

Schedule

Understanding the magnetic properties of nanoparticles is a central issue in the magnetic materials. The finite size effects dominate the magnetic properties of nanoparticles, and become more important as the particle size decreases because of the competition between surface magnetic properties and core magnetic properties. Therefore, in the first half-year, we would focus on the preparation and characterization of the Co nanowires on the GaAs surface by the lithography techniques. In the second half-year, we use the Brillouin light scattering spectroscopy to investigate the magnetic anisotropy of nanostructures by varying the aspect ratio of metallic nanowires and the growth conditions of the Co nanowires.

References

[1] T. Mewes, H. Nembach, J. Fassbender, and B. Hillebrands, Preparation and magnetic anisotropies of epitaxial Co(110) films on MgO(110) substrates, In: *Annual Report 2001*, B. Hillebrands (Ed.), pp. 57 (2001).

[2] J. Fassbender, G. Güntherodt, C. Mathieu, B. Hillebrands, R. Jungblut, J. Kohlhepp, M. T. Johnson, D. J. Roberts, and G. A. Gehring, *Phys. Rev. B*, 57, pp. 5870 (1998).

[3] A. R. Frank, J. Jorzick, M. Rickart, M. Bauer, J. Fassbender, S. O. Demokritov, and B. Hillebrands, Magnetic anisotropies of epitaxial Fe films on vicinal to Ag(001)/GaAs(001) substrates, In: *Annual Report 1999*, B. Hillebrands (Ed.), pp. 48 (1999).

[4] M. Rickart, A. R. Frank, B. F. P. Roos, J. Jorzick, Ch. Krämer, S. O. Demokritov, and B. Hillebrands, Growth and magnetic anisotropies of epitaxial Fe films on vicinal to Ag(001)/MgO(001) substrates, In: *Annual Report 1999*, B. Hillebrands (Ed.), pp. 52 (1999).

Research Proposal: Example 3

Title: Study of the Growth and the Optical Properties of the Self-Organized Nanostructures

Aim

Our purposes are to study the growth of self-organized nanostructures, study optical properties of organized nanostructures, and fabrication of nanodevices.

Background

Self-assembly is an attractive alternative to nanolithography because of high throughput and low cost. Self-assembled nanostructures are expected to show intrinsic bistability in the transport characteristics that can be exploited to realize a non-volatile quantum dot memory also signatures of Coulomb blockade and Coulomb staircase (at room temperature) in quantum dots and wires, which holds out the promise that these structures may find applications in single-electron transistors and other novel nanodevices.

Project Outline

The systems will be investigated with the view to achieving self-organizing low dimensional structures. The first system is to grow self-organized quantum dots on conducting substrate. Our plan is to grow a range of self-organized semiconductor quantum dots primarily by electric field–assisted growth. The growth work will be guided by the optical and transport measurements. The second system that has to be investigated is based on template-assisted growth of nanowires. Using this technique, it is possible to produce one-dimensional organized array of nanowires with well-defined diameters chosen within the range 7 nm to 150 nm oriented perpendicular to the substrate. The plan is to grow porous alumina membrane with long-range order. These porous alumina membranes will then be used to grow one-dimensional nanowires. The nanowires can also be released from the matrix by dissolving the membrane.

As the aim is to grow electrochemically self-organized quantum dots, fully understanding the fundamental scientific principles relating to the mechanism for self-organization of quantum dots is important. It is evident that for most optical device applications, high density of uniform and defect-free quantum dots is required. The use of strain to produce self-organized quantum dot structures has now become a well-accepted approach and is widely used in semiconductors and other material systems. During heteroepitaxial growth of lattice-mismatched materials, strain-induced coherent relaxation occurs and dislocation-free islands are formed that are in the nanometer range in size. This mechanism has been demonstrated for evaporated and electrodeposited semiconductors. Due to quantum size confinement, quantum dots can show atom-like discrete single-electron energy levels. Therefore they may be used as building blocks in single-electron or single-photon devices. However, one of the key problems that need to be solved is the control of their energy level spectrum. Besides the materials electronic structure, this is determined by the size and shape of the quantum dot. We will study the effect of the crystal size for a range of quantum dots, e.g., PbS, PbSe, etc, which show strong size quantization. Finally, the quantum dots exhibit optical properties that are dependent upon the size of the QD itself. Size-dependent optical properties will be investigated as well.

Methods

The nanorods will be grown oriented perpendicular to the plane of the substrate. This could provide viable alternative to the use of nanolithography in producing quantum size effects in one-dimensional systems. The work will be targeted at optical, structural, and transport properties. Vertically oriented well-ordered nanowires can be achieved by template-assisted bottom-up growth using porous anodic alumina (PAA). After investigating technologically important materials, efforts will be made to fabricate the devices out of these structures. The critical concept in our approach is that the self-organized array will allow nanoscale hybrid device functions to be realized. A variety of nanostructures can be created using AAO templates by depositing metals, semiconductors, organics, or their combinations into the pores. The innovation lies in the electrochemical setup of the process to grow the quantum dots and nanowires in a highly uniform manner, which in turn self-organize into regular dimensionalities.

Schedule

The self-organized quantum dots will be carried out in the first half-year. In the second half-year, we will focus on nanowire array, following by the organized nanodevices.

Research Proposal: Example 4

Title: Study of the Self-Assembled Organic Monolayers on Metal Substrate Using Raman Spectroscopy

Aim

Since the electrical studies of organic films require metal contacts, our purpose is to use surface-enhanced Raman scattering (SERS) as a sensitive probe to investigate the structural, electronic, and chemical properties of the interface of the self-assembled monolayers (SAMs) of the complex organic molecules on Pt-coatings. The features of the organic/metal interfaces of our samples are properly varied, in order to achieve surface enhancement of the Raman response of the ultrathin molecular layers.

Background

Raman spectroscopy is applied to probe the structural, electronic, and chemical properties of the ordered self-assembled molecular films on the metallic electrodes. The enhancement of the Raman scattering stems from the contribution of two main mechanisms: electromagnetic (EM) enhancement and charge transfer (CT) enhancement. The EM mechanism involves the giant enhancement of the electromagnetic field near the metal surface because of plasmon excitations in the metal clusters induced by the incident radiation. A charge transfer between the metal and the molecule or vice versa causes the CT effect, commonly refereed as a "first layer effect". This effect strongly enhances the vibrational modes of the molecules having direct contact with the metal. Lots of researches concerning the SERS have been developed for the complex organic molecules SAMs on noble metals. Gold, copper, and silver have been mainly studied in great detail. Haifeng et al. [1–3] have shown that the Raman spectra have been obtained to examine the characteristics of the Inositol hexaphosphate molecules SAMs on copper. Maryt et al. [2] revealed that the complete reduction of the terminal 2-methyl-1,4-naphthoquinone in the SAMs on gold was confirmed by Fourier-transform surface-enhanced Raman spectroscopy. The electrochemical properties of the SAMs aaAzoC4SH and AzoC4SH on a gold surface by XRD and Raman spectroscopy were studied by Zhang et al. [4]. Giuseppe [5] analyzed the surface-enhanced Raman spectra as a function of the monolayer thickness of SAMs on silver by changing the distance in a nanometric way. However, research works in Raman spectroscopy techniques are limited to the study of the electronic properties of the SAMs on gold, copper, and silver surface. As SERS responses depend on the nanostructures of the metal substrate, in this study we would study the organic SAMs on platinum films as a new nanomaterial and characterized by Raman spectroscopy, and we expect that the Raman spectroscopy could confirm the free electrons attributed by platinum surface.

Project Outline

The systems would be investigated with the view to achieving self-assembly organic monolayers on metal surface. The first system is to grow SAMs of organic molecules on Pt substrate. Surface-enhanced Raman spectroscopy is then employed for the characterization of organic molecules-Pt surfaces due to the different orientation of the molecular plane of Pt in thinner layers compared to that in thicker layers. The complex organic interlayer thickness dependence would be observed by SER spectra as well.

Methods

The emphasis of molecular electronics is on synthesizing new organic molecules with different electronic functionalities and on devising new methods to measure the electrical properties of the complex molecules. In order to obtain surface enhancement of the Raman signal, we deposit the organic SAMs on Pt substrate. Pt substrate is prepared by evaporation of Pt film on a glass microscope slide. The head group forms the chemical bond with surface atoms of substrate leading to the pinning of surfactant molecule to the substrate, while the Alkyl interchain van der Waals interactions could assist in formation of ordered molecular structure. The top of the SAMs is used to measure their electrical properties. Imaging the surface topography though AFM, SEM, or TEM can access the quality and uniformity of a monolayer.

Combined surface-enhanced Raman spectroscopy (SERS) and tip-enhanced Raman spectroscopy (TERS) experiments for SAMs on Pt surfaces would be introduced. STM tip-induced frequency shifts for the vibrational modes are observed. This effect is attributed to a distinct weight of the individual contributions to TERS, depending on the particular field distribution.

Schedule

The self-assembled complex organic molecules on Pt surface will be carried out in the first half-year. In the second half-year, we will focus on spectroscopy and microscopy tool.

References

[1] Haifeng Yang, Yu Yanga, Yunhui Yang, Hong Liu, Zongrang Zhang, Guoli Shen, and Ruqin Yu, *Analytica Chimica Acta,* 458 (2005), pp. 159–165.

[2] Maryt Kazemekait, Arunas Bulovas, Zita Talaikyt, Eugenijus Butkus, Vilma Railait, Gediminas Niaura, Algirdas Palaima, and Valdemaras Razumas, *Tetrahedron Letters*, 45 (2004), pp. 3551–3555.

[3] Haifeng Yang, Yu Yang, Zhimin Liu, Zongrang Zhang, Guoli Shen, and Ruqin Yu, *Surface Science,* 551 (2004), pp. 1–8.

[4] Wen-Wei Zhang, Xiao-Ming Ren, Hai-Fang Li, Chang-Sheng Lu, Chuan-Jiang Hu, Hui-Zhen Zhu, and Qing-Jin Meng, *Journal of Colloid and Interface Science,* 255 (2002), pp. 150–157.

[5] Giuseppe Compagnini, Angela D. Bonis, and Rosario S. Cataliotti, *Materials Science and Engineering C*, 15 (2001), pp. 37–39.

[6] Umapada Pal, Jose Garcia-Serrano, Gildardo Casarrubias-Segura, Naoto Koshizaki, Takeshi Sasaki, and Sasaki Terahuchi, *Solar Energy Materials and Solar Cells*, 81 (2004), pp. 339–348.

Research Proposal: Example 5

Title: Nonlinear Optical Study of Metallic Nanoparticles

Aim

Our purpose is to study the second-harmonic light generated from a diffraction grating of metallic nanoparticles with planar inversion symmetry. By measuring the angular distribution of second-harmonic light, we expect to examine the effect in which the diffraction pattern of the grating is superimposed on the intrinsic second-harmonic radiation pattern of the metallic nanoparticles.

Background

Second-harmonic generation (SHG) has been used for over two decades as an optical probe of electronic properties of metal nanoparticles of varying shape, size, composition, and spatial organization. The objectives of metal nanoparticles studies have ranged from measuring electron dephasing to gain insight into the origin of surface-enhanced Raman scattering and assessing the potential for plasmonic applications such as all-optical switching. It is taken for granted that symmetry forbids the generation of second-harmonic light in centrosymmetric nanoparticle systems. Even when asymmetric nanoparticles are arranged so that the overall array has inversion symmetry, SHG is completely suppressed along the illumination direction. This quenching of SHG along the illumination direction holds true for both surface and bulk SHG contributions for this reason, the potential of SHG for probing electron dynamics in metal nanoparticles has been discounted. Recently, it has been proposed that arranging asymmetric nanoparticles in a diffraction grating should provide spatial separation of nanoparticle-generated SH light from both the incident fundamental beam and the substrate-generated SH light. However, asymmetric nanoparticles are not strictly necessary.

Project Outline

The diffracted SHG from metallic nanowires of planar symmetry would be aligned in a symmetric two-dimensional grating, even when optically excited in a symmetric manner. The resulting SH diffraction pattern is unique in that the SHG intensity would increase with diffracted order for a single array and with increasing the angle of observation from the normal. The SHG expect to be dependent on the resonant enhancement between the particle plasmon resonance and the excitation frequency. In this study, we would focus on angles and compare it to the illumination direction. This is a measurement of diffracted SH from nanoparticles with such a high degree of symmetry.

Methods

Nanowire arrays would be fabricated on ITO-coated glass by ion-beam lithography and thermal evaporation. Metal chunks are evaporated over the polymer mask to a given thickness, measured in situ with a quartz crystal microbalance. Atomic force microscopy is used to examine array integrity and nanoparticle structure. We use nanowires because the different axial lengths in a wire give rise to different resonance energies; hence we might probe SHG on- or off-resonance by rotating the array relative to the incident light. It is also easy to control the resonance energy through wire length.

The nanoparticle arrays are illuminated by a mode-locked Nd:YAG laser of 532 nm light. Residual green pump light is blocked with a color filter. Power fluctuations are monitored by a photodiode. Pulse duration is measured with an autocorrelator. The fluence is sufficiently low that SHG is achieved without modifying nanoparticle morphology. The linear extinction coefficients along the major and minor axes of the nanowires are determined separately with a white-light source and rotatable linear polarizer. A focal lens focuses the fundamental beam to a 30–50 micron diameter spot. The arrays are aligned so that the incident polarization pointed along a grating axis. The nanowires point either along the grating axis or at a 45° angle to it. The detector arm rotates in the plane defined by the fundamental propagation and polarization directions. The detector optical train consists of the removable filters, a beamsplitter cube, and a PMT connected to a photon counting module. For PMT measurements, the SH is filtered by a monochromator assuring the spectral purity of the signal. At each observation angle, the PMT signal is optimized; where multiple measurements are acquired at an angle, the highest recorded value is plotted.

Milestones

The fabrication of metallic nanowire arrays on ITO will be carried out in the first half-year. The microstucture of nanowire arrays will be characterized though AFM observation. The optical measurement will be performed in the second half-year.

Research Proposal: Example 6

Title: Photoelectron Spectroscopic Study of the Metalloporphyrins on the Semiconductor Substrate

Aim

Since the electrical properties of the metalloporphyrin require the metallic contacts, our purpose is to use the in situ ultraviolet photoelectron spectroscopy (UPS) as a sensitive probe to investigate the surface electronic structure of the interface of the metalloporphyrins on the semiconductor substrates. The features of the porphyrins/metallic interfaces of our samples are properly varied, in order to achieve the kinetic energy distribution of electrons photoemitted from the ultrathin molecular layers. In addition, these materials would expect to exhibit the magnetic characterization on the Co nanostructures. The Co thickness dependence and the growth temperature dependence on the electrical properties of the porphyrins would be investigated as well.

Background

I. *Photoelectron Spectroscopy*

While one have focused on the use of the electron spectroscopy to investigate the core levels of the surface species, many techniques can provide information on the more weakly bound or less localized valence electronic states. Since this work interests in investigating the surface band structure of the clean metallic surfaces and the ordered self-assembled molecular films, the ultraviolet photoelectron spectroscopy (UPS) is applied to probe the surface valence band. This technique is conceptually identical to X-ray photoelectron spectroscopy (XPS) except that the incident photons are in the range of 20–150 eV. However, there are three significant benefits for surface studies. First, the universal curve of mean free path guarantees that UPS photoelectrons originate from the surface region. Second, the valence band photo-cross-section is large at UPS excitation energies. Third, the energy resolution is excellent because typical laboratory line sources (HeI (21.2eV) and HeII (40.8eV) resonance lamps) have natural linewidth three orders of magnitude smaller than laboratory X-ray sources. Using these lower photon energies of normally less than 40 eV and typically 21 eV, it is clear that only valence levels are accessible. Additionally, the UPS spectrum is more surfaces sensitive than the XPS results [1].

II. *Porphyrins*

Molecular based electronics and photonics devices have attracted great interest in the past decade because the control of electrons or photons at the molecular or supramolecular level is the fundamental process in those electronics and photonics devices. Porphyrin derivatives and those self-organized assemblies are considered as the most likely candidates for such molecular devices; thus many multi-porphyrin systems have been synthesized through coordination interaction as models of the electronics materials and photonics materials. Lots of researches have been developed to study on the porphyrins. Rudolf has studied the adsorption of μ-oxo-iron-meso-tetramethoxyphenylporphyrin on silver, gold, and glassy carbon electrode by using the surface-enhanced Raman scattering [2]. Takeo et al. have studied the migration and transfer of excitation energy in tetraphenylporphyrin monolayers prepared from copolymers by using the fluorescence spectroscopy [3]. Masahiro et al. have used atomic force microscopy (AFM) techniques to observe the self-assembled structures composed of the porphyrin wires on highly oriented pyrolytic graphite substrate [4]. Other porphyrins could be formed on the gold and copper surface using the scanning tunneling microscopy (STM) technique [5,6].

III. *Ferromagnetism*

The study of the magnetic metallic nanostructures has contributed a wide range of new phenomena. The 3-d metals are considered as the ferromagnetism in which the magnitude of the magnetic moment does rely on the electron density. Additionally the electron number changes the Fermi level variation of the magnetization. These have drawn attention of the scientists involved in the research of the basic aspects of low dimensional systems and the materials scientists who intend to apply magnetic nanostructures into devices. The hybrid ferromagnet-semiconductor structures offer several unique benefits to the magnetoelectronic applications. Magnetoelectronics is a new field that is devoted to the invention and development of electrical device structures that incorporate a ferromagnetic element. Generally, the magnetic anisotropy is a fundamental property in determining the magnetic behavior of thin and ultrathin film magnets. From a theoretical point of view, an anisotropy plays an important role in stabilizing ferromagnetism in 2-d magnets.

The magnetic anisotropies have been observed in a wide range of ultrathin magnetic films made of the 3-d metals. Mewes et al. [7] have studied the magnetic anisotropies of the Co films on MgO substrates and revealed that the resulting magnetic behavior is the same high level as the Co films grown on the Cu substrates proposed by Fassbender et al. [8]. Frank et al. have shown the uniaxial magnetic anisotropy contribution to the Fe films on the Ag(001)/GaAs(001) substrates [9]. Rickart et al. have studied the magnetic anisotropies of the Fe films on the Ag(001)/MgO(001) substrates [10].

These results have led us to extend investigation of Co-porphyrins on the semiconductor substrate because it has the possibility of the dilute magnetic semiconductors (DMSs) as they combine the two fields in the condensed matter physics: magnetism and semiconductor. DMSs are the key material of the spintronics that applies the correlation between charge and spin of electrons to devices. The surface effects in the metallic nanoparticles are of great important because the electrical properties of the nanoparticles are strongly influenced when their surface is in contact with different media. Consequently, the proof that Co-porphyrins films on the GaAs substrates are the DMSs would give a large contribution to the spintronics because the Curie temperature of the existing ferromagnetism DMSs, such as the Mn films on the GaAs substrates, is too low to be used for the spintronics devices. To elucidate the electronic origin of the Co-porphyrins on the GaAs surface, in situ UPS would be introduced for this study in order to observe the electrical properties in the Co-porphyrins deposited onto the GaAs substrates. The relation between the electronic structure and the Co nanoparticle characteristics (size-dependent) would be investigated. Additionally, as the UPS responses depend on the structures of the substrate, in this research, we would also study the Co-porphyrins on the GaAs substrate instead of other existing noble metal (gold, silver, and copper, for example), and we expect that the UPS could confirm the free electrons attributed by the semiconductor surface.

Project Outline

The systems would be investigated with the view to achieving self-assembly porphyrins monolayers on metallic surface. The first system is to grow SAMs of porphyrins on the GaAs substrates by using molecular beam technique in the ultrahigh vacuum (UHV) conditions approximately 1×10^{-10} mbar at 500–600 K. Then Co is deposited onto the porphyrin layer by the shadow deposition techniques. The in situ UPS is then employed for the characterization of the complex organic molecules-GaAs surfaces due to the different orientation of the molecular plane of GaAs in thinner layers compared to that in thicker layers. The UV spectra would observe the complex organic interlayer thickness dependence as well. On the other hand, the Co-porphyrins deposited onto the GaAs substrates would be shown a transition of the magnetic anisotropy depending on its thickness. In situ UPS would then investigate the thickness and the temperature dependence of magnetism affecting the electrical properties.

We would show that the photoelectron spectroscopic tool as UPS might apply to the ferromagnetic deposited on the self-assembled monolayers of the large organic molecules as porphyrins adsorbed on the GaAs surface. The UPS sensitivity depends on the ordering of the substrate (density of the GaAs films and density of aggregates) and on the nature of the adsorbed molecules. This is also directly related to the quality of the ferromagnetic and the SAMs formation on the GaAs wafer. Furthermore, when the intermediate used to fix the GaAs substrate is correctly chosen, the photoemission spectra could be then detected. Moreover, the features of the UPS spectrum expect to depend on the classical interference phenomenon between the s–d interband activity of the GaAs surface and the molecular properties of the adsorbate. Finally, we would estimate the order inside the molecular layer deposited on the GaAs surface.

Methods

The emphasis of molecular electronics is on synthesizing the complex organic molecules with different electronic functionalities and on devising new methods to measure the electrical properties of the complex molecules. In order to obtain the UPS signal, we deposit the metalloporphyrins on the GaAs substrates by using the microfabrication techniques developed in the field of semiconductor physics. The nanoscale V-shaped groove structures are first constructed on the surface of the GaAs wafer; 100–200 nm pitch with 10 mm × 10 mm patterned area of V-groove shape. The (100) oriented GaAs wafer is masked by photoresist and then a stripe pattern and is printed by using electron beam lithography. By wet etching rate, V-shaped grooves are formed. The porphyrin thickness ranging from 10 nm to 100 nm at each growth temperature is measured by a quartz crystal oscillator, which is calibrated by an atomic force microscope (AFM) to confirm the quality and uniformity of a monolayer. After such a microstructured substrate is available, the head group of porphyrins forms the chemical bond with surface atoms of substrate leading to the pinning of surfactant molecule to the substrate, while the Alkyl interchain van der Waals interactions could assist in formation of ordered molecular structure. The Co nanowires would be then prepared by evaporation of Co film by using the molecular beam lithography (EBL) machine in the ultrahigh vacuum on the porphyrin layer in the multichamber molecular beam epitaxy (MBE) system. The Co is deposited onto the porphyrin monolayers at the flux angle with respect to the template normal direction. The low mobility of Co atoms on the monolayers enables one to make Co nanowires as thin as several tens of nanometers. The temperature for the Co-porphyrins deposited onto the GaAs substrate is kept constant during the deposition. The process of the growth is confirmed by taking the in situ low energy electron diffraction (LEED) images. It is well known that changes in the GaAs substrate preparation and growth conditions could cause significant changes in the electrical properties of the Co-porphyrins. In order to carry out the thickness dependent electronic measurements without having such impact, the Co-porphyrins are prepared exactly on the same GaAs substrate under identical growth conditions. Then in situ UPS checks the electronic structure of the Co-porphyrins. The experimental setup is explained elsewhere [11].

Schedule

Understanding the electrical properties of nanoparticles is a central issue in the electronic materials. In the first half-year, we would focus on the preparation and characterization of the Co deposited onto the self-assembled complex organic molecules on the GaAs surface by the shadow deposition and the lithography techniques, respectively. In the second half-year, we use the UPS to investigate the electrical properties of the metalloporphyrins by varying the aspect ratio of metallic films and the growth conditions of the Co nanowires. The finite size effects dominate the magnetic properties of Co nanoparticles, and become more important as the particle size decreases because of the competition between surface magnetic properties and core magnetic properties, leading to the changes of their electrical properties.

References

[1] Gerken F., Foldstrom A. S., Barth J., Johansson L. I., and Kunz C., *Phys. Scr.,* 32 (1985), pp. 43.

[2] Rudolf H., *Electrochemical Acta,* 33 (1988), pp. 1619–1627.

[3] Takeo O. and Shinzaburo I., *Thin Solid Films,* 500 (2006), pp. 289–295.

[4] Masahiro K., Hiroaki O., Hirofumi T., and Takuji O., *Thin Solid Films,* 499 (2006), pp. 23–28.

[5] Terui T., Yokoyama S., Suzuki H., Mashiko S., Sakurai M., and Moriwaki T., *Thin Solid Films,* 499 (2006), pp. 157–160.

[6] Kamikado T., Sekiguchi T., Yokoyama S., Wakayama Y., and Mashiko S., *Thin Solid Films,* 499 (2006), pp. 329–332.

[7] Mewes T., Nembach H., Fassbender J., and Hillebrands B., In: *Annual Report 2001*, B. Hillebrands (Ed.), pp. 57 (2001).

[8] Fassbender J., Güntherodt G., Mathieu C., Hillebrands B., Jungblut R., Kohlhepp J., Johnson M. T., Roberts D. J., and Gehring G. A., *Phys. Rev. B,* 57 (1998), pp. 5870.

[9] Frank A. R., Jorzick J., Rickart M., Bauer M., Fassbender J., Demokritov S. O., and Hillebrands B., In: *Annual Report 1999*, B. Hillebrands (Ed.), pp. 48 (1999).

[10] Rickart M., Frank A. R., Roos B. F. P., Jorzick J., Krämer C. H., Demokritov S. O., and Hillebrands B., In: *Annual Report 1999*, B. Hillebrands (Ed.), pp. 52 (1999).

[11] Gottfried J. M., Flechtner K., Kretschmann A., Lukasczyk T., Steinrück H.-P., *JACS* (2006) (Unpublished).

Research Proposal: Example 7

Title: Structures and Magnetic Properties of Cu Nanowires by XRD Technique

Aim

Since the ordered arrays of metallic nanowires are expected to play an essential role as materials for interconnects and high-density magnetic storage devices because of their unique electrical and magnetic properties, one promising technique for the integration of nanowires into well-defined architectures is their deposition into ordered templates. Our purpose is to synthesize the arrays of copper nanowires by electrodeposition in porous templates as anodic aluminum oxide (AAO). The atomic model and magnetic properties of these nanoparticles would be studied by means of XRD then.

Background

The contribution from XRD to the subject has recently been enhanced by diffraction instruments at synchrotron sources where they are supplied with X-rays that are very bright, highly polarized, and tunable in energy. All these attributes are a help when it comes to measuring weak reactions in a diffraction pattern due to the relatively few electrons in valence states, which possess angular anisotropy, and possibly a magnetic moment. The importance of observations made by analyzing the weak reactions can hardly be exaggerated, for the electrons in question participate in a host of physical properties, including covalency, superconductivity, magnetoresistance, ferro- and pyroelectricity, structural phase transitions, and all manner of magnetic phenomena. However, a successful framework for analyzing the weak features in a diffraction pattern gathered on a spatially ordered material is based on an atomic model. One of the attractive features of this model is that the atomic quantities, such as spin and orbital magnetic moments and orbital quadrupole moments, can be probed by XRD experimental technique.

Project Outline

The systems will be investigated with the view to achieving self-organizing low dimensional structures. The system that has to be investigated is based on template-assisted growth of nanowires. Using this technique, it is possible to produce the organized array of nanowires with well-defined diameters chosen within the range 7 nm to 150 nm oriented perpendicular to the substrate. The plan is to grow porous alumina membrane with long-range order. These porous alumina membranes will then be used to grow nanowires. The nanowires can also be released from the matrix by dissolving the membrane.

On the basis of atomic models, one can add X-ray Bragg diffraction in which the intensity of a reaction is increased by tuning the X-ray energy to an atomic resonance, although the suitability of an atomic model needs to be tested case by case. Resonance enhancement increases the visibility of certain features in a diffraction pattern, and it provides an element selectivity. Intensities can be sensitive to polarization in the primary and secondary X-ray beams, and this sensitivity is another useful aspect of resonant Bragg diffraction. The wavelength of the X-rays is determined by the energy of the resonance event being exploited and very often the wavelength is too long to satisfy the Bragg condition at more than two, or three, reactions. The resonant Bragg diffraction would provide a sensitivity at the level of a small fraction of an electron, and data gathered can be used to infer the wave function of the valence state which accepts the photo-ejected core electron. We would review observations made by analyzing diffraction measurements in terms of an atomic model, with some emphasis on X-ray resonant Bragg diffraction.

As the aim is to grow electrochemically self-organized quantum wires, fully understanding the fundamental scientific principles relating to the mechanism for self-organization of quantum wires is important. It is evident that for most optical device applications, high density of uniform and defect-free quantum wires is required. The use of strain to produce self-organized quantum wire structures has now became a well-accepted approach and is widely used in semiconductors and metallic material systems. During heteroepitaxial growth of lattice-mismatched materials, strain-induced coherent relaxation occurs and dislocation-free islands are formed that are in the nanometer range in size. This mechanism has been demonstrated for evaporated and electrodeposited metallic materials. Due to quantum size confinement, quantum wires can show atom-like discrete single-electron energy levels. Therefore they may be used as building blocks in single-electron or single-photon devices. However, one of the key problems that need to be solved is the control of their energy level spectrum. Besides the materials electronic structure, this is determined by the size and shape of the quantum wires. We will study the effect of the crystal size for a range of quantum wires, which show strong size quantization. Finally, the quantum wires exhibits magnetic properties that are dependent upon the size of the QW itself. Size-dependent magnetic properties will be investigated as well.

Methods

The nanowires will be grown oriented perpendicular to the plane of the substrate. This could provide viable alternative to the use of nanolithography in producing quantum size effects. The work will be targeted at structural and transport properties. Vertically oriented well-ordered nanowires can be achieved by template-assisted bottom-up growth using porous anodic alumina (PAA). A variety of nanostructures can be created using AAO templates by depositing metals, semiconductors, organics, or their combinations into the pores. The innovation lies in the electrochemical setup of the process to grow nanowires in a highly uniform manner, which in turn self-organize into regular dimensionalities. XRD spectra would be recorded at room temperature in a diffractrometer with Cu Kα radiation and a graphite monochromator, while Mg Kα line is used as X-ray source.

Milestones

The self-organized quantum wires will be carried out in the first year. In the second year, we will analyze the crystal structure and magnetic properties on nanowire array though diffraction patterns.

8.4 Problems

Pb.1. Show the difference between oral and poster presentations.

Pb.2. Select one paper of your interest and present it in terms of oral or poster format.

Pb.3. Write a new research proposal when you need to get the financial support from your affiliation.

Pb.4. Read the following research papers and present them in terms of the presentation format: oral or poster presentations and/ or research proposal.

Research Paper (1)

Abstract

In this article, we simply design the solar tracking system and construct a solar collector system for year 2015 in Bangkok, Thailand. The analytical model is calculated via altitudes and azimuth angles of the sun. Our experimental result is in well agreement with the calculation in terms of altitude and azimuth. This solar tracking system is therefore applied to a dish solar collector showing the thermal energy of 961.69 W at a maximum temperature of 543.3 K with a maximum electric power of 3.395 W from our thermoelectric modules.

Introduction

The solar trackers mostly use a sensor module [1–5] or the sun position control systems [6,7]; however, there are some benefits and drawbacks for these systems. That is, although the sensor module offers a good precision for solar tracking, it does not work well with poor weather condition. On the other hand, the sun position control system is independent of the weather condition, but its accuracy is still severe [6,7]. Herein we introduce for the first time the Analema curve into the calculation of the sun position in order to get better accuracy.

In details, a heat collection from solar radiation by tracking of the sun is dependent on the position between the earth and the sun in a day of year [8]. We are able to use at least two parameters, altitude and azimuth angles, to identify the position of the sun [9]. These factors may vary according to a location of latitude of the observer [10]. In this paper we pay attention to determine the altitude and azimuth in Bangkok time zone at our institute for year 2015. Firstly, we calculate the position of the sun by using National Oceanic and Atmospheric Administration (NOAA) equation of time, depending on period, earth's rotation angle on its axis, longitude, and latitude. Then, the calculation data will be used to design and construct a sundial and a solar parabolic dish [11]. Since the errors from the obtained altitude and azimuth angle as compared with the theoretical calculation are not over ±2°, this method is suitable for automatic sun-tracking system. Furthermore, we accomplish a good thermal process from this method. We expect that our new homemade instrument is very useful to convert the heat from the sun to be electricity. A combination of light sensor module with calculation of earth–sun position is also expected to solve problems of both weather condition and instrument accuracy, our new design and construction will be present in the next paper.

Design and Fabrication

We can in principle calculate the position of the sun in the horizon system in terms of altitude and azimuth angles as shown in Fig. 1 [12]. On one hand, altitude A_α or the angle of the sun up from the horizontal is written as $A_\alpha = \sin^{-1}(\sin\delta\sin\ell_{\phi W}^{E} + \cos\delta\cos\omega\cos\phi)$, where $\ell_{\phi W}^{E}$ is latitude angle ranging from 0° at the Equator to 90° (North or South) at the poles.

Declination angle δ is written as $\sin\delta = \cos\left[(n-173)\dfrac{180°}{182.6}\right]$ $\sin(23.45°)$ in which n is a date sequence of the year. For example, $n = 1$ represents 1-Jan-2015. Hour angle ω in hours is written as $\omega = 15(T_{AST} - 12)$, whilst an apparent solar time in hours is defined as $T_{AST} = t_{LCT} \pm \dfrac{t_{Zw}^{E} \pm \ell_{\lambda w}^{E}}{15} + \dfrac{t_{EQT}}{60} - D$. t_{LCT} is local time in hours and t_{Zw}^{E} is time zone (7 for Bangkok, Thailand). $\ell_{\lambda w}^{E}$ is longitude line of the observer, and D is daylight saving time. National Oceanic and Atmospheric Administration (NOAA) equation of time t_{EQT} in minutes can be calculated by $t_{EQT} = 60 \sum_{k=0}^{5} \left[A_k \cos\left(\frac{360kn}{365.25}\right) + B_k \sin\left(\frac{360kn}{365.25}\right) \right]$ and

coefficients A_k and B_k are shown in Table 1 [8].

Table 1 Coefficients A_k and B_k

k	A_k (hr)	B_k (hr)
0	2.0870×10^{-4}	0
1	9.2869×10^3	-1.224×10^{-1}
2	-5.2258×10^{-2}	-1.5698×10^{-1}
3	-1.3077×10^{-3}	-5.1602×10^{-3}
4	-2.1867×10^{-3}	-2.9823×10^{-3}
5	-1.5100×10^{-4}	-2.3463×10^{-4}

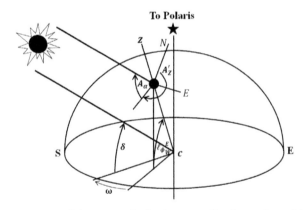

Figure 1 Position of the sun in the horizon coordination system. C is earth's center.

On the other hand, azimuth A_Z or the angle between a reference position (North or South) and a line from the observer to the sun projected on the same plane is defined as $\cos\omega \geq \left(\frac{\tan\delta}{\tan\phi}\right)$, $A_Z = 360° + A_Z'$ or $\cos\omega <$

$\left(\frac{\tan\delta}{\tan\phi}\right)$, $A_Z = 360° + A_Z'$ in which $A_Z' = \sin^{-1}\left(\frac{-\cos\delta\sin\omega}{\cos\alpha}\right)$.

Next, we construct sundial and solar concentrator system to collect data of altitude and azimuth angles from 1-Jan-2015 to 31-Dec-2015 (06:00–18:00). On one hand, we design and construct the steel rods perpendicular to the sundial as shown in Fig. 2. On the other hand, we design and construct the solar collector dish [12–14] and receiver served as the solar concentrator system as shown in Fig. 3.

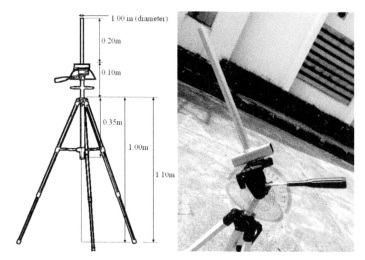

Figure 2 Sundial.

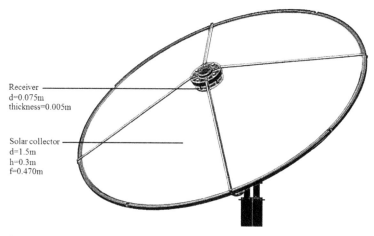

Figure 3 Solar collector dish and receiver.

Focal length of the dish is calculated by $f = \dfrac{d^2}{16h}$. d and h are diameter and height of the dish, respectively [15]. In this experiment, we use d = 1.5 m and h = 0.3 m, and we obtain f = 0.470 m. Receiver (diameter = 0.075 m, thickness = 0.005 m) is installed at a focus point of the dish in order to collect all the heat from the solar radiation. The thermal energy is therefore calculated by $Q_u = I_b A_a \eta_0 - A_r \left[h(T_r - T_a) + \varepsilon\sigma(T_r^4 - T_a^4) \right]$, in which I_b is the light intensity (W/m^2) and A_a is the solar collector dish area (m^2). η_0 is combined light efficiency of the solar concentrator and A_r is receiver area (m^2). h is the convection coefficient of the solar concentrator (W/m^{2}K) and T_r is the temperature of (Cu) the receiver. T_a is the environmental temperature (K) and ε is the emissivity of the solar concentrator. σ is Stefan's constant.

Results and Discussion

The experiment is carried out at our institute (King Mongkut's Institute of Technology Ladkrabang, Bangkok, Thailand) located at latitude ($\ell^E_{\phi w}$ = 13°43′N) and longitude ($\ell^E_{\lambda w}$ = 100°47′E). The measurement time is 06:00–18:00 from 1-Jan-2015 to 31-Dec-2015. According to the above-mentioned observer, the earth–sun position visibly causes a change between altitude and azimuth. Declination angle and the equation of time for year 2015 indicate the altitude and they are shown in Fig. 4. The compensation time between the solar motion and local time for year 2015 is calculated from amplitude of the equation of time (−0.5633 minutes). This indicator tells us the declination angle of the earth's axis in one cycle around the sun.

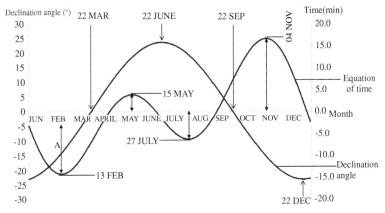

Figure 4 Equation of time and declination angle from 1-Jan-2015 to 31-Dec-2015. A is amplitude.

In details, we find that the declination angle (−0.25°) is still the same on 22-Mar-2015 and 22-Sep-2015 due to the earth's axis perpendicular to solar radiation. In the meantime, we see that the declination angle is still the same on 22-Jun-2015 (23.44°) and 22-Dec-2015 (−23.44°). The former aspect is caused by the earth's axis close to the sun and the later one is attributed to the earth's axis far away the sun.

The experimental data from both altitude and azimuth fit well with calculation data as shown in Fig. 5. The errors between them are less than ±2°. Therefore, these data are suitable for constructing the Analemma curve as shown in Fig. 8.

It is noted that the Analemma curve in Fig. 8 is fit by using altitude–azimuth angles in Fig. 6 and equation of time-declination angles in Fig. 7 [16].

Finally, the position of the sun from the calculation can be applied to collect the solar radiation via our solar concentrator system as shown in Fig. 9.

We find that our receiver stores the thermal energy very well (thermal energy = 961.69 W) at maximum temperature T_{max} of 543.3 K at 13:30. By using our thermoelectric modules, we find that a maximum electric power P_{max} is 3.395 W at 13:50. It is noted that we are unable to measure P_{max} at T_{max} due to limitation of the thermoelectric modules.

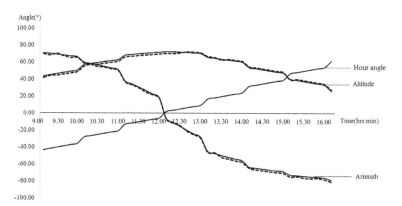

Figure 5 Altitude and azimuth angles from calculation and experiment (dashed and solid lines stand for experimental and calculation data, respectively), as well as hour angle on 13-Mar-2015 between 09:00 and 16:00.

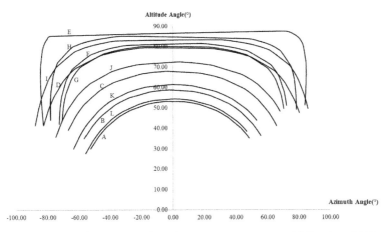

Figure 6   Altitude and azimuth angles. A to L represent Jan-2015 to Dec-2015.

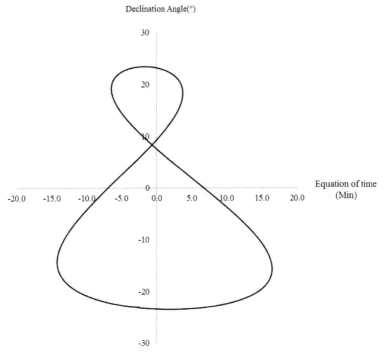

Figure 7   Analemma curve at 13:30.

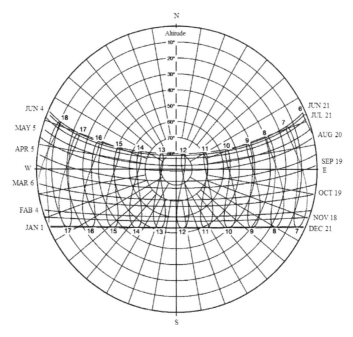

Figure 8 Analemma at latitude = 13°45′N (BKK).

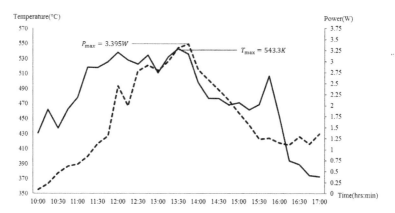

Figure 9 Temperature as a function of time (hrs:min) of the receiver as shown in solid line and the electric power as a function of time (hrs:min) from thermoelectric modules as shown in dash line.

Conclusion

We design and construct the solar tracking system to determine the position of the solar motion at our institute (King Mongkut's Institute of

Technology Ladkrabang, Bangkok, Thailand) with latitude of 13°43′N and longitude of 100°47′E for year 2015. The experimental data from both altitude and azimuth fit well with the calculated one. These data are suitable for constructing the Analemma curve. Identified position of the sun motion is then applied to the solar concentrator system. It is found that our receiver not only provides very high thermal energy at a maximum temperature of 543.3 K, but also offers a maximum electric power of 3.395 W. This thermal process is therefore expected to offer a great potential for conversion into electric power in near future.

References

[1] M. T. A. Khan, S. M. S. Tanzil, R. Rahman, and S. M. S. Alam, Design and construction of an automatic solar tracking system, *6th Int. Conf. Elec. Compt. Eng. (ICECE)* (2010), pp. 326–329.

[2] U. K. Okpeki and S. O. Otuagoma, Design and construction of a bi-directional solar tracking, *Int. J. Eng. Sci.*, 2(5) (2013), pp. 32–38.

[3] Y. C. Park and Y. H. Kang, Design and implementation of two axes sun tracking system for the parabolic dish concentrator, *ISES2001 Solar World Congress* (2001), pp. 749–760.

[4] X. Jin, G. Xu, R. Zhou, X. Luo, and Y. Quan, A sun tracking system design for a large dish solar concentrator, *Clean Coal Energy*, 2(2B) (2013), pp. 16–30.

[5] P. Roth, A. Georgiev, and H. Boudinov, Design and construction of a system for sun-tracking, *Renew. Energy*, 29(3) (2004), pp. 393–402.

[6] Y. Rizal, S. H. Wibowo, and Feriyadi, Application of solar position algorithm for sun-tracking system, *Energy. Proc.*, 32 (2013), pp. 160–165.

[7] M. Mirdanies, Astronomy algorithm simulation for two degrees of freedom of solar tracking mechanism using C language, *Energy Proc.*, 68 (2015), pp. 60–67.

[8] L. Morison, *Introduction to Astronomy and Cosmology*, John Wiley & Sons, London, 2008.

[9] S. Ray, Calculation of sun position and tracking the path of sun for a particular geographical location, *Int. J. Emer. Technol. Adv. Eng.*, 2(9) (2012), pp. 81–84.

[10] W. B. Stine and R. W. Harrigan, *Solar Energy Fundamentals and Design: With Computer Applications*, John Wiley & Sons, New York, 1985.

[11] G. Prinsloo and R. Dobson, *Solar Tracking*, Prinsloo, South Africa, 2014.

[12] J. A. Duffie and W. A. Beckman, *Solar Engineering of Thermal Processes*, John Wiley, New York, 2013.

[13] N. D. Kaushika and K. S. Reddy, Performance of a low cost solar paraboloidal dish steam generating system, *Energy Conversion Mgt.*, 41 (2000), pp. 713–726.

[14] R. Y. Nuwayhid, F. Mrad, and R. Abu-Said, The realization of a simple solar tracking concentrator for university research applications, *Renew. Energy*, 24(2) (2001), pp. 207–222.

[15] S. Pairoj, A parabolic solar concentrator with different receiver materials, Master Thesis, King Mongkut's Institute of Technology Ladkrabang, Thailand.

[16] P. Lynch, The equation of time and the Analemma, *Bull. Irish Math. Soc.*, 69 (2012), pp. 47–56.

Research Paper (2)

Abstract

Recently the removal of trace element using biodegradable polymers is important. This paper involves the preparation and evaluation of chitosan/ polyethylene glycol blend served as a heavy metal removal system. The author has prepared the various blending system with different composition ratios and different cross-linking density. Experimental results indicate that the swelling degree and thermal property of the blend film are correlated with blend ratio and crosslink density. The blend film is then investigated for its metal-binding performance. Copper sorption capacity is one of the major potential applications in a field of wastewater treatment.

Introduction

Despite a number of researches on metal chelating by chitosan (CS) or other polymers as in [1–5], no study can be found on any development of the metal-chelating capacity by the chitosan/polymer blend system. The blending strategy is an important approach for biodegradable polymers, not only to improve their inferior properties of the structural components, but also to achieve the efficient absorbent characteristics in order to treat wastewater and to recover the trace elements [6–10]. Therefore, the author has prepared and studied the blending property of glycol/chitosan (CS/PEG) film. The potential application of blended CS/PEG film is applied for metal chelating. Also, the biodegradabilities of CS and PEG are usually allowed. The differential scanning calorimeter (DSC) will be utilized to evaluate the miscibility of the blend. If blended polymers are immiscible, different phases will be easily detected. Finally, the removal of heavy metal ion throughout the blend products will be investigated their potentials served as a new copper (II) ion absorbing agent.

Experiment

Polymer Blend. Chitosan powder (CS with molecular weight = 480,000 Da and degree of deacetylation = 75–85%) and polyethylene glycol (PEG with molecular weight = 6,000) powder including ethylene glycol diglycidyl ether (EGDGE) with different blend ratio were dissolved in acetic acid solution using magnetic stirrer for 72 hrs. It was noted that polyethylene glycol was inactive for the cross-linking with chitosan and we used EGDGE as a cross-linker. The blend solution was subsequently degassed to remove air bubbles, and then spread onto a Teflon dish. Afterwards, the gel was dehydrated. It was noted that the acid film was neutralized by base solution. The CS to PEG blend ratios were 1:4, 1:2, 1:1, 2:1, and 4:1 by weight. Ethylene glycol diglycidyl ether (EGDGE) was used to each sample in different quantity (0.5 mL, 1.5 mL, and 3.0 mL). Form of polymer blend used in this experiment was the film.

Swelling Degree. Dried film was soaked in deionized water until reaching equilibrium state. The obtained wet film was wiped in order to separate the excess water on the surface of the film, and weighed. The level of swelling was expressed in accordance with the weight of swollen sample per weight of dried sample.

Thermal Property. Thermal property of 10 mg dried film was measured by using a Perkin–Elmer Pyris DSC-7. Scanning range was from −100°C to 200°C.

Metal Absorption. Metal stock solution served as synthetic wastewater was prepared by dissolving an amount of the known metal salts ($CuCl_2$) in DI water. The pH of the test solution was controlled to 4–5 using hydrochloric acid. The obtained solution was then adsorbed with the blend film until reaching equilibrium absorption, and the mixture became sediment. The clear liquid was separated by filtration. Amount of Cu (II) ions remaining in this solution was verified by titration with the standard sodium thiosulphate solution [11–14]. The amount of metal absorption was, therefore, calculated by subtracting the amount of copper from the initial amount and reported as (mg) per gm basis of the blend film.

Results and Discussion

Film Characterization. Blend solution is homogeneous and transparent, changing from colorless to slightly yellowish as the chitosan (CS) content increases (Fig. 1). The obtained blend film is translucent. It becomes soft when it is immersed in water. These results indicate that intermolecular hydrogen bonds, in case of a low polyethylene glycol (PEG) composition, are possibly formed between hydroxyl groups, or amino groups of CS molecule and ether groups of PEG molecule, whilst intramolecular hydrogen bonds are formed in CS chain at a high PEG composition. It is also reliable for the cross-linking formation through a reaction of amino, hydroxyl groups of CS and carboxyl groups of cross-linking agent—ethylene glycol diglycidyl ether (EGDGE)—in the presence of PEG.

Figure 1 CS to PEG blend ratios were 1:4 (left) and 4:1 (right).

Swelling Degree. Swelling degree tends to increase when EGDGE is decreased (Fig. 2). This is because decreasing crosslink density increases the spacing or phase domain size of the polymer network. High crosslink density leads the chain segmental immobility to produce its bigger free space. On the other hand, the swelling degree is correlated with the proportion of CS in the blend film. This is because the amorphous phase from CS disturbs the crystallinity of the blend to lose the crystalline structure in the overall structure. This irregular structure leads the polymer chains to pack loosely. This structure also facilitates the efficient water diffusion into the polymer networks. In the contrary, when we increase the crystallinity of the blend by PEG content, water molecules weaken in penetrating into more compact structure.

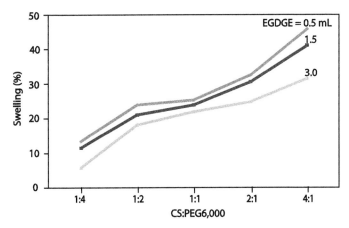

Figure 2   Swelling degree of the blend films.

Thermal Characterization. In Fig. 3, the DSC output spectra of pure CS do not indicate any endothermic transition over the experimental temperature region. This is attributed to the lack of any crystalline or any phase changes during the heating process. The same figure also exhibits the DSC scan of pure PEG, and the endothermic peak centered around 70°C is employed to the thermal degradation of PEG.

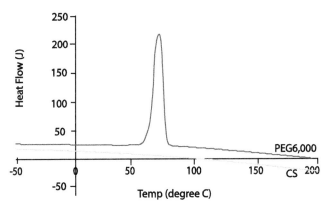

Figure 3   DSC thermograms of CS and PEG6,000.

The DSC scans (Figs. 4 and 5) of the observed blend film reveal that only one endothermic peak is present for each blend film. The temperature of the endothermic peak ranging from 50°C to 70°C corresponds to the melting temperature (T_m) of PEG, and in agreement with the results suggested by Huh et al. [15] and Zeng et al. [16]. Before the endothermic peak of T_m, a slope change (−35°C to −45°C) is observed attributing to the glass transition temperature (T_g) of the blend film.

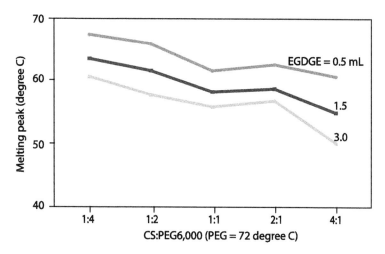

Figure 4   Melting peak of the blend.

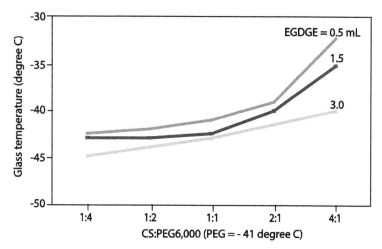

Figure 5   Glass transition temperature of the blend.

Melting temperature (T_m) of the blend film is shifted toward the low temperature when the CS content is increased (Fig. 4). The enthalpy changes in terms of heat flow, of course, controlled by the chemical structure of the blend are conformed to T_m (not shown). The CS hinders the crystallinity of the blend, both in the reduction of the size of crystalline and broadening the distribution of the crystalline. On the other hand, cross-linking agent involves the exchange between a van der Waals bond for a shorter covalent bond and more compact bonds. This results in a further decrease in T_m.

Conversely, glass temperature (T_g) of the blend film appears to increase when we increase the amorphous CS phase via the presence of its molecular flexibility and weak interchain force (Fig. 5). The T_g is decelerated by cross-linking agent because not all the carboxylic groups of cross-linker are involved in cross-linkage, but the remaining ones might have more strong hydrogen bonds.

Absorption Capacity. In Table 1, the CS content in the blend film is obviously shown to have a strong ability to absorb the metal ion.

Table 1 Average value of the absorption capacity in mg copper per gm of the blend film

Materials	0.5 mL EGDGE	1.0 mL EGDGE	1.5 mL EGDGE
CS:PEG = 1:4	60 mg/g	40 mg/g	38 mg/g
CS:PEG = 1:2	70 mg/g	50 mg/g	40 mg/g
CS:PEG = 1:1	75 mg/g	55 mg/g	53 mg/g
CS:PEG = 2:1	78 mg/g	62 mg/g	60 mg/g
CS:PEG = 4:1	80 mg/g	75 mg/g	70 mg/g

CS, PEG, and EGDGE are chitosan, polyethylene glycol, and ethylene glycol diglycidyl ether, respectively.

The presence of mainly NH groups of CS molecule is involved in metal-chelating process via the dominant mechanisms as follows:

$$-NH_3^+ = -NH_2 + H^+ \tag{1}$$

$$-NH_2 + Cu^{2+} + 2OH^- = [CuNH_2(OH)_2] \tag{2}$$

Amount of cross-linking agent is inversely correlated with adsorption ability of the blend film. The low content of cross-linker easily unattacks on the characteristic groups ($-NH_2$) of CS backbone, and, at the same time, unreacts with ether group of PEG molecule. The blend film with the smallest content of cross-linker also shows the best swelling performance of water as previously described in the swelling degree. Therefore, metal ion is easily combined with the polymer matrix. Since more water is able to be absorbed in the blend and water might stimulate the diffusion of copper ion, the copper absorption rate is enhanced. For further adsorption study, the theoretical models such as isotherm model and kinetic model will be present in the next paper.

From all results purposed, it is clear that the structure of the blend film can be considered as a potential alternative of a new composite adsorbent for copper ion waste streams.

Conclusion

A modified polyethylene glycol/chitosan (PEG/CS) blend film was prepared with the help of a cross-linking agent and studied for its properties. The blend film with lowering of the crosslinked epoxide system attempts to impart a high swelling degree. The polymer swelling suppresses the native crystallinity of the blend, increasing the accessibility of water molecules to the sorption sites when we increase the CS constituent. In thermal analysis, CS and cross-linker are found to have significance in the crystallinity of the blend. Moreover, the polymer blend is suitable for the chelation of copper ion, which is chosen to demonstrate the utility of such blend on the metal complex. The results reveal that the copper absorption capacity of the blend film is correlated with the CS content, but inversely correlated with the amount of cross-linking agent. We believe that the concept of modifying the metal-chelating capacity of this blend film will suggest a wide range of applications in the metal ion sorptions.

References

[1] K. Vathsala, T. V. Venkatesha, B. M. Praveen, and K. O. Nayana, Electrochemical generation of Zn-chitosan composite coating on mild steel and its corrosion studies, *ENG.*, 2(8) (2010), pp. 580–584.

[2] B. Samiey, C. H. Cheng, and J. Wu, Organic–inorganic hybrid polymers as adsorbents for removal of heavy metal ions from solutions, *Materials*, 7 (2014), pp. 673–726.

[3] S. E. Cahyaningrum, N. Herdyastusi, and D. K. Maharani, Immobilization of glucose isomerase in surface-modified chitosan gel beads, *Res. J. Pharm. Biol. Chem. Sci.*, 5(2) (2014), pp. 104–111.

[4] B. K. Ahn, D. W. Lee, J. N. Israelachvili, and H. Waite, Surface-initiated self-healing of polymers in aqueous media, *Nature Mat.*, 13 (2014), pp. 867–872.

[5] Z. K. George and N. B. Dimitrios, Recent modifications of chitosan for adsorption applications: A critical and systematic review, *Mar. Drugs*, 13(1) (2015), pp. 312–337.

[6] R. Jozef, J. Eva, N. Marta, K. Radek, and E. Tomas, The toxic effect of chitosan/metal-impregnated textile to synanthropic mites, *Pest Mgnt. Sci.*, 69(6) (2013), pp. 722–726.

[7] V. Mohanasrinivasan, M. Mishra, J. S. Paliwal, S. K. Singh, E. Selvarajan, V. Suganthi, and C. Subathra Devi, Studies on heavy metal removal efficiency and antibacterial activity of chitosan prepared from shrimp shell waste, *3 Biotech*, 4(2) (2014), pp. 167–175.

[8] J. Wang and C. Chen, Chitosan-based biosorbents: Modification and application for biosorption of heavy metals and radionuclides, *Bioresources Technol.*, 160 (2014), pp. 129–141.

[9] A. Yasabie and S. Omprakash, Removal of chromium by biosorption method (chitosan), *Int. Lett. Nat. Sci.*, 3 (2014), pp. 44–55.

[10] S. Idris, K. Murat, A. Gulsin, B. Talat, and C. Talip, Preparation and characterisation of biodegradable pollen–chitosan microcapsules and its application in heavy metal removal, *Bioresources Technol.*, 177 (2015), pp. 1–7.

[11] H. O. Triebold, *Quantitative Analysis*, Read Books, New York, 2007.

[12] W. T. Elwell and I. R. Scholes, *Analysis of Copper and Its Alloys*, Elsevier, New York, 2013.

[13] I. P. Alimarin and M. N. Petrikova, *Inorganic Ultramicroanalysis*, Elsevier, New York, 2013.

[14] G. L. Heath, *The Analysis of Copper and Its Ores and Alloys*, BiblioLife, New York, 2015.

[15] K. M. Huh, Y. W. Cho, H. Chung, I. C. Kwon, S. Y. Jeong, T. Ooya, W. K. Lee, S. Sasaki, and N. Yui, Supramolecular hydrogel formation based on inclusion complexation between poly(ethylene glycol)-modified chitosan and α-cyclodextrin, *Macromol. Biosci.*, 4(2) (2004), pp. 92–99.

[16] M. Zeng, Z. Fang, and C. Xu, Effect of compatibility on the structure of the microporous membrane prepared by selective dissolution of chitosan/synthetic polymer blend membrane, *J. Membr. Sci.*, 230 (2004), pp. 175–181.

Research Paper (3)

Abstract

We have designed and constructed a brand new dish solar tracking system. This hybrid system is composed of a sensor module and a local clock time equation systems, and they all work together whatever the weather conditions are. Experimental results from this system show that the tolerances of the azimuth and altitude angles are not over ±2° showing a high accuracy of the system when we have performed this system in Bangkok, Thailand. Moreover, the measured temperature from the solar receiver is 508.25 K. This system is then expected to work well with the solar collector for the electric power conversion in future.

Introduction

Nowadays the solar tracking systems have 2 ways. We can use a sensor module [1–4] or we can apply NOAA equation to calculate a position of the sun [5,6]. Both of them have some disadvantages when the weather is unpredictable. One module usually consists of 4 optical sensors for 4 directions (North, South, East, and West) [7]. If one of these sensors malfunctions, we are unable to continue to track the sun's position. The drawback of NOAA equation is that there are some errors caused by the movement line of the sun for each month in a year. To overcome these problems, we have designed and generated a new hybrid solar tracking system. We have combined new sensor module and new local clock time equation systems into one system. In details, we use eight optical sensors from LDR (light-dependent resistor) for four directions (North, South, East, and West). This means that we detect the sun position with three sensors for one direction. In this case, if one of three sensors malfunctions, our system is still working. We will also show that our new hybrid solar tracking system has a good accuracy in tracking the movement of the sun and we can avoid the weather problem. In addition, our system can be installed in any place by just input new latitude and longitude data of the location of interest.

Design and Construction

New solar tracking system is composed of a sensor module and a local clock time equation systems as well as an atmospheric sensor. On one hand, if the weather is clear, a sensor module system will detect the light. On the other hand, if the weather is not clear, the sensor module will stop and a local clock time equation system will run instead. It is noted that the encoder is mounted at two DC servo motors to check azimuth and altitude angles in order to identify the position of the sun [8].

Sensor Module. Optical sensor is LDR (light-dependent resistor) served as a light detector with the aperture angle of 60°. As we detect the sun position with three sensors for one direction, this allows us to increase the exposure of the sun up to 180°. The LDR is designed according to a principle of shadow from the aluminum bar as shown in Fig. 1(a). For altitude-LDR as shown in Fig. 1(b), when *the sun's rays are parallel with the aluminum bar in altitude line*, three sensors at the upper position represented by $T_1 T_2 T_3$ and another three sensors at the lower position represented by $T_4 T_5 T_6$ can detect the maximum light intensity. Our tracking program is set at "ON" whereas a logic state is set to be "1". For azimuth-LDR as shown in Fig. 1(b), three sensors at the left position represented by $Z_1 Z_2 Z_3$ and another three sensors at the right position represented by $Z_4 Z_2 Z_6$ cannot detect the light. Our tracking program is set at "OFF" position whereas a logic state is set to be "0". By contrast, if *the sun's rays are parallel with the aluminum bar in azimuth line*, the altitude-LDR will show "OFF" and "0", whilst the azimuth-LDR will show "ON" and "1".

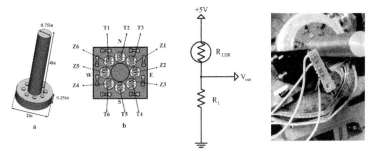

Figure 1 (a) Sensor module (LDR type). (b) Altitude-LDR (top and bottom) and azimuth-LDR. (c) Atmospheric sensor.

Local Clock Time Equation. As the sun goes along the altitude and azimuth lines as shown in Fig. 2 at the angle range of 0° to 180°, we introduce the following equation to find a position of the sun motion for each day.

$$SP = \frac{T_{LCT}}{180°} \tag{1}$$

where SP is a position of the sun (min/degree) and T_{LCT} is local clock time (min).

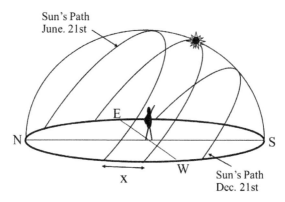

Figure 2   Sun motion for each month of a year.

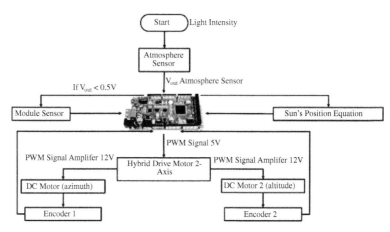

Figure 3   Diagram of microcontroller.

When the sun moves along the altitude and azimuth lines, we can identify a moving distance of the sun for each month in a year (represented by x as shown in Fig. 2) by using NOAA equation [9]. This moving distance is transferred to a microcontroller as shown in Fig. 3.

Atmospheric Sensor. Atmospheric sensor as shown in Fig. 1(c) is LDR served as a weather detector relying on a principle of voltage divider according to Eq. (2). The measured voltage (V_{out}) as shown in Fig. 1(c) is transferred to a microcontroller to convert analog to digital signals. Therefore, we apply V_{out} in digital mode to edit the sensor module and the local clock time systems.

$$V_{out} = 5 \times \frac{R_{LDR}}{R_{LDR} + R_1} \; ; \; \left(\begin{array}{ll} V_{out} < 0.5 \text{ V} & \text{(Sensor Module)} \\ 0.5 \text{ V} < V_{out} < 2 \text{ V} & \text{(Local Clock Time Equation)} \end{array} \right) \qquad (2)$$

Here R_{LDR} is LDR resistance and R_1 is resistance's constant (1 kΩ).

Construction. Mechanical structure of solar tracking device as shown in Fig. 4 includes the mechanical parts of azimuth and altitude lines and they are summarized in Table 1.

Table 1 Mechanical structure of solar tracking system

Mechanical part	Motor	Encoder
Altitude line	12 V, 5 rpm	12 V, 600 CPR
Azimuth line	12 V, 100 rpm	12 V,100 CPR

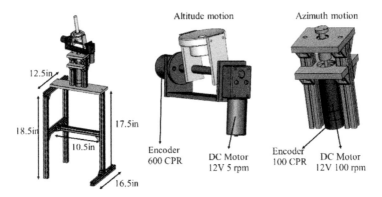

Figure 4 Mechanical structure.

We check the default setting for the solar tracking system. Next compass and azimuth lines are set at 0° indicating the North as shown in Fig. 5 in order to obtain a less tolerance of azimuth and altitude angles as much as we can. After that, we test the accuracy of solar tracking system.

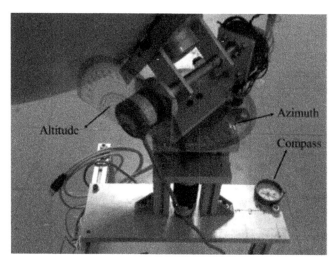

Figure 5   Angle measurement of altitude and azimuth lines.

In order to collect all the heat from the solar radiation, we design and construct the dish solar collector at a focal length of 0.21 m and receiver (diameter = 0.006 m, length = 0.0015 m, material: brass) as shown in Figs. 6 and 7 [10]. Then we measure the outdoor temperature from this receiver to show the performance of our solar tracking system.

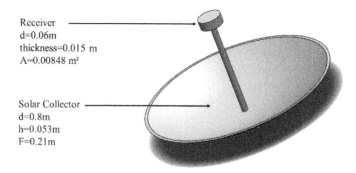

Receiver
d=0.06m
thickness=0.015 m
A=0.00848 m²

Solar Collector
d=0.8m
h=0.053m
F=0.21m

Figure 6   Design of dish solar collector and receiver.

Results and Discussion

We install dish solar collector and receiver as well as thermocouple as shown in Fig. 7. Also, we install the hybrid solar tracking device at our King Mongkut's Institute of Technology Ladkrabang, Bangkok,

Thailand. This place is located at latitude $\ell_{\phi W}^{E} = 13°43'N$ and longitude $\ell_{\lambda w}^{E} = 100°47'E$. We test the accuracy and performance of the solar tracking system at 09:00–16:00. The average outdoor temperature is 33–40°C. Altitude and azimuth angles as well as the output voltage are measured and plotted in Fig. 8.

Figure 7 Construction of dish solar collector and receiver.

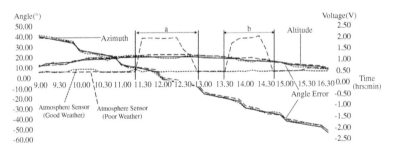

Figure 8 Measured altitude and azimuth angles on 10-Jan-2016 (short-dashed lines) and 13-Jan-2016 (long-dashed lines). Calculated altitude and azimuth angles (solid lines). Atmospheric sensor gives the output voltage.

The errors from the measured altitude and azimuth angles are calculated as follows:

$$Azimuth\ Angle\ Error = NOAA_{AZ} - LCT_{AZ} - SM_{AZ}$$

$$Azimuth\ Angle\ Error = NOAA_{AL} - LCT_{AL} - SM_{AL} \qquad (3)$$

where NOAA_{AZ} and NOAA_{AL} are altitude and azimuth angles, respectively, calculated by NOAA equation. LCT_{AZ} and LCT_{AL} are altitude and azimuth angles, respectively, calculated by local clock time equation. SM_{AZ} and SM_{AL} are altitude and azimuth angles, respectively, measured by the sensor module system.

By combining sensor module and the local clock time equation systems with atmospheric sensor, we can tract the sun position in any weather conditions. In a case of a clear weather, V_{out} will be less than 0.5 V (short-dash line in Fig. 8). LCT_{AZ} and LCT_{AL} in Eq. (3) are then equal to zero. In this case, the sensor module system will work, whereas the local clock time equation system will stop. On the other hand, if the weather is not clear, V_{out} will be in between 0.5 V and 2 V (long-dash line; *a* and *b* in Fig. 8). We find that SM_{AZ} and SM_{AL} in Eq. (3) are equal to zero. In this case, the sensor module system will stop and the local clock time equation system will work instead. This allows us to obtain a high accuracy of tracking system with tolerances of altitude and azimuth angles of less than ±2°. Last, we check the performance of the system. That is, we check the temperature from the receiver and we measure the light intensity from Pyranometer. It is found that the temperature and light intensity based on 10-Jan-2016 at 13:00 are 235.25°C and 901.96 W-m^2, respectively.

Conclusion

We have designed and constructed the new hybrid solar tracking system at King Mongkut's Institute of Technology Ladkrabang, Bangkok, Thailand located at 13°43′N latitude and 100°47′E longitude. In this system, we have combined the new sensor module and the local time equation systems into one system. Our findings indicate that the tolerances of the azimuth and altitude angles are less than ±2° showing a high precision of the tracking system. This system can be installed anywhere by just inputting new latitude and longitude points. Furthermore, the temperature from the heat detector is 508.25 K showing that it will be a useful system because it can work together with the solar collector in order to produce electricity in future.

References

[1] P. Saha and S. Goswami: *Int. J. Emerg. Technol. Adv. Eng.*, 4 (2014), pp. 286.

[2] A. Sarhan, A. E. Wakeel, and A. B. Kotb: *Aerospace Sci. Avia. Technol.*, 13 (2009), pp. 1–12.

[3] A. J. Alzubaidi: *Int. J. Comput. Technol. Appl.*, 6 (2015), pp. 295–302.

[4] T. C. Cheng, W. C. Hung, and T. H. Fang: *Int. J. Photoenergy*, 2013 (2013), pp. 1–7.

[5] S. Ray: *Int. J. Emerg. Technol. Adv. Eng.*, 2 (2012), pp. 81–84.

[6] M. Mirdanies: *Int. Conf. Sustainable Energ. Eng. Appl.*, 68 (2015), pp. 60–67.

[7] P. Roth, A. Georgiev, and H. Boudinov: *Renew. Energy,* 29 (2004), pp. 393–402.

[8] G. Prinsloo and R. Dobson: *Solar Tracking* (Prinsloo, South Africa, 2014).

[9] Information on http://www.esrl.noaa.gov/gmd/grad/solcalc/azel.html

[10] S. Lekchaum and K. Locharoenrat: *Appl. Mech. Mater.* (2016), in press.

Research Paper (4)

Abstract

In this article, we have designed and fabricated the gamma-type Stirling engine based on thermodynamics' concept. This engine is attached onto a parabolic dish of a solar collector. The engine shows a good performance in terms of compression ratio, external work, total pressure, and engine speed. Our engine offers the thermal efficiency of 30.59% in order to reach the output mechanical power of 0.934. The temperature difference of 137 K can maintain very well the heat collection in the solar collector even when the weather conditions are poor. Furthermore, our materials are environmentally friendly, and this design is expected to be in the applications for the solar tracker in the future.

Introduction

Currently there are two devices (solar panel and solar concentrator) for the solar collection to convert heat to electricity. Although the life-times of these devices are not so different, solar concentrator quite works well in terms of the high thermal efficiency as compared with solar panel [1]. In solar concentrator, the solar receiver is generally used at the focal point of the parabolic dish. Alternatively, we are able to use Stirling engine in place of the receiver. This system can also be applied to the solar tracking system [2]. The principle of the Stirling engine relies on the temperature difference between the cylindrical displacer (hot side) and power cylinder (cold side). We can easily apply the cylindrical displacer in place of the receiver. The important parameters of the engine consist of displacer length, compression ratio, total pressure, thermal efficiency, output mechanical power, and Beale number. Herein we introduce a new design and construction of the gamma-type Stirling engine relying on the working temperature difference caused by the heat source of the solar concentrator. Our entire engine is made of brass that has a good thermal conductivity as compared with other works [3–4]. We also use air served as a working fluid in place of inert gas [3–7] in the entire engine system that is environmentally friendly. We will show the performance of the engine in terms of the thermal efficiency and the output mechanical power. We expect that our engine will be useful for the heat collection in future.

Design

We firstly design a gamma-type Stirling engine. The configuration, including the cross-section of the engine, is displayed in Fig. 1. Based on thermodynamic principle, as the expansion–compression process exists in one cycle of the engine, we obtain the geometrical parameters and the performance of the engine as follows [8].

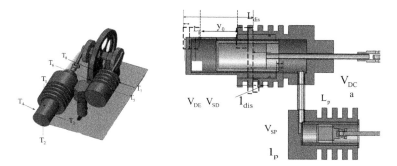

Figure 1   Configuration (left) and cross-section (right) of Gamma-type Stirling engine.

Expansion volume V_E is calculated as

$$V_E = \frac{V_{SD}}{2}(1 - \cos\theta) + V_{DE} \tag{1}$$

V_{SD} is a swept volume of a displacer piston (cm^3), θ is a crank-shaft angle, and $V_{DE} = \dfrac{\pi D_{dis}^2 L}{4}$.

Compression volume V_C is written as

$$V_C = \frac{V_{SP}}{2}\left[1 - \cos(\theta - \varphi)\right] + V_{DC} \tag{2}$$

V_{SP} is a swept volume of a power piston (cm^3), $\varphi = \pi/2$, is a phase angle between a displacer piston and a power piston.

$$V_{DC} = \frac{1}{4}(\pi D_{dis}^2 [L_{dis} - (L + y_0 + l_{dis})] + \pi D_p^2 [L_p - (l_p + y_p + a)])$$

Total pressure P in one cycle of the engine is written as

$$P = \frac{P_{mean}\sqrt{1 - \lambda^2}}{1 - \lambda \cos(\theta - \delta)} \tag{3}$$

Here P_{mean} = 1.013 bars is the average pressure of the engine. λ is the ratio between A and B in which

$$A = (\gamma^2 + 2(\gamma - 1)\chi_s \cos\varphi + \chi_s^2 - 2\gamma + 1)^{1/2}$$

and $B = (\gamma + 2\gamma\upsilon + \chi_s + 2\varsigma + 1)$.

$\gamma = T_C/T_H$ in which T_H (K) is the temperature at hot side and T_C (K) is the temperature at cold side. $\chi_s = V_{SP}/V_{SD}$ is the swept volume ratio, $\upsilon = V_{DE}/V_{SD}$ and $\varsigma = V_{DC}/V_{SD}$ are the expansion and compression volume ratios, respectively. $\delta = \tan^{-1}\left(\dfrac{\chi_s \sin\varphi}{\gamma + \cos\varphi + 1}\right)$.

Compression ratio ξ in one cycle of the engine is calculated by

$$\xi = 1 + (V_{SD}/V_{DE}). \tag{4}$$

Thermal efficiency η in one cycle of the engine is defined as

$$\eta = 1 - \gamma. \tag{5}$$

Output mechanical power P_s is defined as

$$P_s = W_i N = \frac{2\pi N\tau}{60} \tag{6}$$

in which $W_i = W_E + W_C$ is the external work done per one cycle (mJ).

$$W_E = \int P dV_E = \frac{P_{mean} V_{SD} \pi \lambda \sin \delta}{1 + \sqrt{1 - \lambda^2}}$$ is the work at the expansion volume.

$$W_C = \int P dV_C = \frac{P_{mean} V_{SD} \pi \gamma \sin \delta}{1 + \sqrt{1 - \lambda^2}}$$ is the work at compression volume. N is the speed of the engine (rpm), $\tau = r_p F \sin \varphi$ is the torque of the engine (N-cm), and r_p is the length of the crank. F is the external force.

Beale number B_o is calculated as [9]

$$B_o = \frac{P_s}{p_{mean} f V_{DS}} \tag{7}$$

Here $f = \dfrac{N}{60}$.

Experiment

The geometrical parameters of the designed engine are constructed. This engine is then mounted onto a parabolic dish of the solar collector (a focal length of 21 cm at a temperature of 508 K for the heat collection). The temperatures at the hot side (T_1, T_2, T_3, T_4) and the cold side (T_5, T_6, T_7, T_8) (as shown in Fig. 1) of the engine are measured by the thermocouples (Model: TM-9475D). The engine speed (N) is detected by using Digital Techometer (Model: DIGICON DT-246L).

Results and Discussion

The performance of the constructed engine is shown in Table 1 in which the lengths of the displacer (L) are varied from 1.00 to 1.40 cm at a constant stoke of 2.00 cm. Hot wall side is kept at T_H = 448 K and cold wall side is kept at T_C = 311 K. Therefore, the temperature difference of 448 – 311 = 137 K provides the thermal efficiency (η) of 30.59%. This temperature difference can help temperature of the solar concentrator not too low when it works at the low temperature conditions. When the length of the displacer (L) is increased at constant phase of 90°, it is found that the external work (W_i), total pressure (P), and output mechanical power (P_s) are increased, whilst the engine speed (N) is reduced.

Beale number (B_o) in Table 1 conforms to Ref. [9]. This allows us to achieve the output mechanical power (P_s) up to 0.934 W at the length of the displacer of 1.40 cm. It is noted that our engine does not work when the length of the displacer is over 1.40 cm. This is a limitation of our engine that needs to further improve in the future.

Table 1 Performance of the Sterling engine

L (cm)	ξ	B_o	V_{DE} (cm^3)	W_i (mJ)	P (kPa)	N (rpm)	P_s (W)
1.00	3.000	0.015	4.98	102.100	140.220	480	0.816
1.20	2.666	0.016	5.89	112.070	142.890	471	0.877
1.40	2.428	0.017	6.87	121.350	145.550	462	0.934

Compression ratio (ξ) of 2.428 (at the length of the displacer of 1.40 cm) is similar to Ref. [8]. This allows us to achieve the output mechanical power (P_s) of 0.934 W, and it can be explained as follows. Compression ratio results in a heat transfer between the displacer and power pistons, resulting in the heat-mechanical conversion. If compression ratio is more than 2.428, a pressure drop caused by aerodynamic friction happens.

It is interesting to observe that when the length of the displacer (L) is increased, compression ratio (ξ) is decreased. It can be explained by Eqs. (1), (2), and (4) that the length of the displacer (L) is proportional to the expansion volume (V_E); however, it is inversely proportional to compression volume (V_C).

Conclusion

We have designed and constructed the gamma-type Stirling engine, in which the length of the displacer is changed from 1.00 to 1.40 cm. This engine is then mounted onto a parabolic dish of the solar collector. The engine shows the improvement of some parameters, for instance, compression ratio, external work, total pressure, and engine speed. Moreover, the engine provides the thermal efficiency of 30.59%. This allows us to achieve the output mechanical power of 0.934. The temperature difference of 137 K can keep well the heat collection in the solar collector in the case of poor weather conditions. Furthermore, our materials are environmentally friendly. This design is also very useful in the applications for the solar tracking system in the future.

References

[1] Information on http://www.geni.org/globalenergy/research/review-and-comparison-of-solar-technologies/Review-   and-Comparison-of-Different-Solar-Technologies.pdf

[2] G. Prinsloo and R. Dobson: *Solar Tracking* (Prinsloo, South Africa, 2014).

[3] M. Hooshang, R. Askari Moghadam, S. Alizadeh Nia, and M. Tale Masouleh: *Renew. Energy*, 74(2015), pp. 855–866.

[4] A. Sripakagorn and C. Srikam: *Renew. Energy*, 36 (2011), pp. 1728–1733.

[5] A. Asnaghi, S. M. Ladjevardi, P. Saleh Izadkhast, and A. H. Kashani: *Int. Scholar. Res. Network*, 2012 (2012), pp. 1–14.

[6] C. Cinar and H. Karabulut: *Renew. Energy*, 30 (2005), pp. 57–66.

[7] A. D. Minassians: *Stirling Engines for Low-Temperature Solar-Thermal-Electric Power Generation* (Berkeley, California, 2007).

[8] G. Walker: *Stirling Engines* (Oxford Clarendon, USA, 1980).

[9] W. Beale, U.S. Patent 3552120 (1971).

Research Paper (5)

Abstract

Solar cell increased efficiency as well as lowered cost per unit of power generated. These developments were of great interest in the last few decades. Dye-sensitized solar cells are one of the solutions to that problem. ZnO is a semiconducting material commonly used to build photovoltaics. It absorbs the sunlight in UV part of the spectrum; hence the light harvesting can be enhanced by the lower-energy part of the spectrum absorbing dye, such as coumarin-153. This work shows the effect of mixing coumarin-153 with ZnO in sol–gel films made by spin-coating technique, as well as stability of those films.

Introduction

Dye-sensitized solar cells (DSSCs) have been a subject of numerous researches as an alternative to silicone photovoltaic since their first discovery in 1991 due to high efficiency of DSSCs, low cost and ease of making process, under ambient conditions [1–3]. DSSCs are composed of metal oxide film, a photosensitive dye, electrolyte-hole transporter, and a counter electrode. In that system, the sensitizer (the dye) plays a vital role. The main function of the dye is to absorb the light in visible and near-IR part of the spectrum, considering that the metal oxide only absorbs in UV part of the spectrum [2, 3].

The need for alternative to ruthenium-based dyes appeared because of high cost of rare metals, difficult process of synthesizing mentioned dyes, as well as complicated purification process [2,4]. On the other hand, organic dyes could be synthesized easily, or even extracted from plants. Structural modifications are as well easily obtained by simple synthesis process, in order to tune position and intensity of the charge transfer transition. Organic dyes also have a higher molar extinction coefficient compared to ruthenium-based dyes [2,5].

The common system among organic metal-free dyes is donor–π-bridge–acceptor (D-π-A), which ensures effective charge transport and separation. Electron donor affects molecular energy level; hence, it affects absorption spectra of the molecule. The other role of the electron-donor component is in geometric structure of the molecule and, by that, providing efficient charge transfer. Red-shifting absorption is achieved with π conjugated component of the dye molecule which affects HOMO/LUMO energy levels [1,6].

Many metal-free organic dyes showed promising results in photovoltaic application. Those are derivatives of coumarin, triphenylamine, carbazole, and indoline [2,4,6]. Coumarin-153 is a derivative of coumarin, with additional electron-donor and electron-acceptor groups. Electron-donor group greatly lowers the energy of $S_{\pi\pi^*}$ and $T_{\pi\pi^*}$. In addition electron-acceptor group is as well shifting fluorescence and absorption maximum peak to longer wavelengths, resulting in absorption peak of coumarin-153 at 430 nm, compared to 365 nm for coumarin [7]. For that reason, goal of this work is to efficiently combine ZnO, which is very commonly used material for solar cells because of its optical properties, with coumarin-153, which reported promising results in photovoltaic application with TiO$_2$, in order to enhance efficiency of ZnO sunlight absorption [7–9].

Experiment

For making sol–gel ZnO/coumarin-153 films, polymethymethacrylate (PMMA) was measured and dissolved in dichloromethane (DCM). Ethanol solution of coumarin-153 (0.01 mol/dm^3) was added, followed by 50 weight % ZnO solution. Different amounts of all the components are presented in Table 1. ZnO films were made with ethanol instead of coumarin-153 ethanol solution, following Table 2. Finally, coumarin-153 film was made by mixing PMMA with coumarin-153 ethanol solution, pure ethanol and DCM, according to Table 3. All cocktails were sonicated in ultrasonic bath for 30 minutes, in order to make a homogeneous solution. Figure 1 shows coumarin-153, ZnO and mix coumarin-153 and ZnO films, respectively.

Table 1 Components for sol–gel ZnO/coumarin-153 films

PMMA [gm]	Coumarin-153 [µL]	ZnO solution [µL]	DCM [µL]
0.2	400	117.5	482.5
0.2	400	129.5	470.5
0.2	400	141	459

Table 2 Components of ZnO films

PMMA [gm]	Ethanol [µL]	ZnO solution [µL]	DCM [µL]
0.2	400	117.5	482.5
0.2	400	129.5	470.5
0.2	400	141	459

Table 3 Components of coumarin-153 films

PMMA [gm]	Coumarin-153 [µL]	Ethanol [µL]	DCM [µL]
0.2	400	100	500

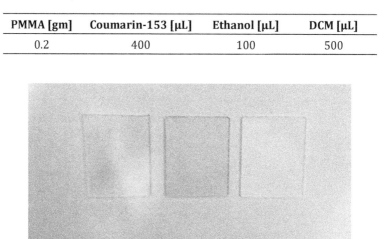

Figure 1 Coumarin-153 (left), ZnO (middle), and coumarin-153 + ZnO films (right), respectively.

After drying, absorbance and fluorescence of the films were measured using the equipment presented at Figs. 2 and 3. As a reference for absorbance, the film was made by dissolving 0.2 gm PMMA in 500 µL DCM and 500 µL ethanol.

Figure 2 The equipment for measuring absorbance: (1) detector, (2, 4) optical cables, (3) sample holder, and (5) Tungsten as a light source.

In order to examine stability of those films, they were exposed to UVB radiation for 180 minutes. Stability was investigated by measuring fluorescence at pre-determined times: 20, 50, 80, 110, 140, and 180 minutes.

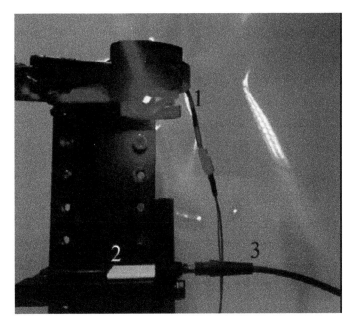

Figure 3 The equipment for measuring fluorescence spectra: (1) UVB light source, (2) the sample, and (3) optical cable (to detector).

Results and Discussion

ZnO is a semiconducting material commonly used in building photovoltaic. However, the need for more economical ways of acquiring energy led to the need of enhancing efficiency of used materials. One way is dye-sensitization. ZnO has a band gap of about 3.37 eV, which means that ZnO absorbs the light in UV part of the spectrum. The dye with lower band gap energy absorbs the light in lower frequencies, transferring electrons to the conductive band of ZnO. That way, higher light utilization is reached [1,7,10].

Figure 4 shows absorbance of different amounts of ZnO mixed with 4×10^{-3} mol/dm^3 coumarin-153. The first peak, at 370 nm, is from ZnO and the second one, at 420 nm, belongs to coumarin-153. It is clearly seen that with increasing amount of ZnO, summed absorbance is also higher.

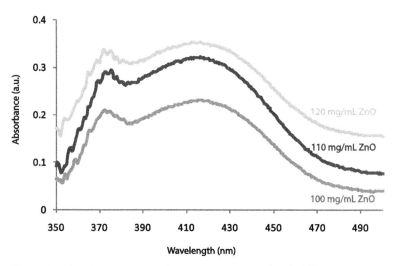

Figure 4 Absorbance spectra of coumarin-153 mixed with different amounts of ZnO.

Figure 5 represents, similarly to Fig. 3, fluorescence of different amounts of ZnO mixed with 4×10^{-3} mol/dm^3 coumarin-153. Obtained data shows that increasing amount of ZnO is inversely proportional to fluorescence, which is to be expected.

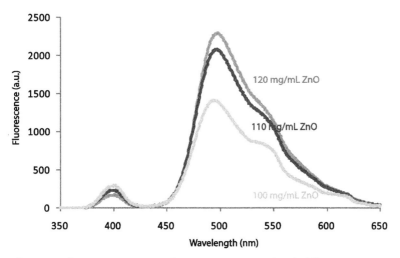

Figure 5   Fluorescence spectra of coumarin-153 mixed with different amounts of ZnO.

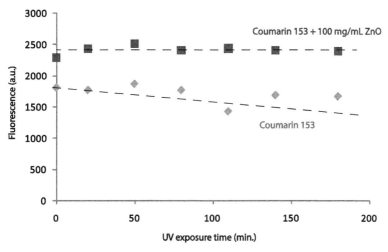

Figure 6   Fluorescence of the films exposed to UVB radiance over time.

The data in Fig. 6 shows fluorescence as a function of the time when the films were exposed to UVB radiance. The coumarin-153 shows lower photo-stability than coumarin-153 + 100 mg/mL ZnO (representative sample), with fluorescence decreasing with time. On the other hand, it is clear from the data obtained that the stability of the mixed complex is a lot higher. That makes this mix suitable for building long-lasting photovoltaic.

Conclusion

With the goal of creating effective and stable photovoltaic, the effect of adding coumarin-153 to ZnO films was investigated. Higher summed absorbance indicates that coumarin-153 has an important role in light harvesting. The dye absorbs the light in higher-frequencies region, transferring its electrons to the conductive band of ZnO, and that way enhancing semiconductive activity of ZnO. The data obtained from the measurement of fluorescence of the film exposed to UVB radiation over time also shows that the dye-sensitizing positively influences the stability of the mixture, compared to the organic dye alone, which is unstable under UVB radiation.

References

[1] C. Zhong, J. Gao, Y. Cui, T. Li, and L. Han, *J. Pow. Sour.*, **273**, 831 (2015).

[2] An. Margalias, K Seintis, M. Z. Yigit, M. Can, D. Sygkridou, V. Giannetas, M. Fakis, and E. Stathatos, *Dyes Pigm.*, **121**, 316 (2015).

[3] Z. Wan, C. Jia, Y. Duan, X. Chen, Y. Lin, and Y. Shi, *Org. El.*, **14**, 2132 (2013).

[4] K. D. Seo, I. T. Choi, Y. G. Park, S Kang, J. Y. Lee, and H. K. Kim, *Dyes Pigm.*, **94**, 469 (2012).

[5] L. Tan, L. Xie, Y. Shen, J. Liu, L. Xiao, D. Kuang, and C. Su, *Dyes Pigm.*, **100**, 269 (2014).

[6] H. Tan, C. Pan, G. Wang, Y. Wu, Y. Zhang, Y. Zou, G. Yu, and M. Zhang, *Org. El.*, **14**, 2795 (2013).

[7] P. C. Ricci, A. Da Pozzo, S. Palmas, F. Muscas, and C. M. Carbonaro, *Chem. Phys. Lett.*, **531**, 160 (2012).

[8] X. Liu, G. Wang, A. Ng, F. Liu, Y. H. Ng, Y. H. Leung, A. B. Djurisic, and W. K. Chan, *Appl. Surf. Sci.*, **357**, 2169 (2015).

[9] X. Yu, X. Yu, J. Zhang, and H. Pan, *Mat. Lett.*, **161**, 624 (2015).

[10] A. Peles, V. P. Pavlovic, S. Filipovic, N. Obradovic, L. Mancic, J. Krstic, M. Mitric, B. Vlahovic, G. Rasic, D. Kosanovic, and V. B. Pavlovic, *J. Alloy Compd.*, **648**, 971 (2015).

Pb.5. Read the following review papers and present them in terms of the presentation format—oral or poster presentations and/or research proposal:

Review Paper (1)

Abstract

Although second harmonic generations from gold and copper nanowires have been the subject of extensive studies, a complete understanding of the mechanism of the enhancement is still missing. The aim of review article is to address these issues by performing systemic studies of mechanism of the resonant enhancements between gold- and copper nanowires in terms of surface reconstruction and relativistic effect, and implement these types of candidate for the potential applications in optical communications and optical devices.

Introduction

In the previous paper, the author selected copper nanowires as a promising candidate for future interconnection applications and measured the wavelength dependence of their second harmonic (SH) intensity [1]. The nanowires of copper are fabricated by the shadow deposition method [1]. The second harmonic generation (SHG) from copper nanowires is then compared with gold nanowires produced by the same fabrication technique [2]. A strong enhancement of SH intensity from gold nanowires is observed at $2\hbar\omega$ = 3.3 eV. For copper nanowires, the author has discovered that at the SH photon energy of 4.4 eV, the SH response is induced by a resonant coupling between the fundamental field and local plasmons in the wires. This localized plasmon excitation is judged to exist for copper nanowires, but it is not observed for SHG from gold nanowires. However, the linear reflectivity spectra of gold nanowires show plasmon absorption for the incident field perpendicular to the wire axes.

Since the position of resonant peak in copper nanowires is found to be different from that of gold nanowires, this review article represents two candidate origins of the observed SH intensity peaks. One is relativistic effect because the atomic nucleus of gold is heavier than that of copper, the important influences of relativity at least for the core electrons are to be expected for gold. The other is surface reconstruction. An increasing tendency for the missing-row structures when going from copper to gold will be discussed.

To our knowledge, optical response of copper and gold nanowires may depend on the history of the generation process of the excited electrons concerning whether the initial state of the electrons originates from the *d*- or *s*-band [3]. By considering the one outer *s*-electron and *d*-electrons of gold and copper lying 2–5 eV below the Fermi energy, the author proposes Fig. 1, the excitation of these free electrons that may generate the SH signal. In Fig. 1(a), the transitions from occupied s state to an unoccupied s state from the same band structure are possible. For this case the electrons must scatter with lattice defects or impurities in order to conserve momentum. With the high density of initial *d*-band states in Fig. 1(b), the excitations of the *d*-band electron to the unoccupied part of the *s*-band are also to be dominant in the SHG.

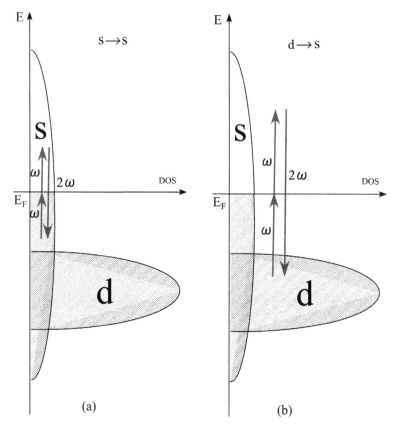

Figure 1 Illustration of SHG process from copper and gold in case (a) the s-electrons are optically excited, whereas in case (b) predominantly d-electrons contribute to the SHG.

Regarding relativistic effect to the initial state of the excited electrons of metals, this effect is important for heavy-atom metal systems producing, in addition to spin-orbital splitting [4–7], orbital contraction and other effects [8–11], which results in changes in spectroscopic and optical properties. As compared with copper, there is more compression of the sp-electrons in gold because its atomic nucleus is heavy enough to cause the relativistic core contraction via the orthogonality constraint for all valence orbitals [12]. This relative shift between d- and s-band causes an increased overlap of their density of states (DOS), which allows for an increase in the sd hybridization and then an increased bonding charge density of the electrons. Figure 2(b) confirms a relativisitically enhanced stronger sd hybridization and d-d interaction in gold than that of nonrelativistic one in Fig. 2(c).

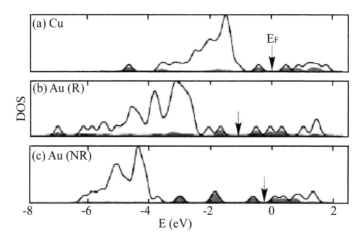

Figure 2 DOS of copper and gold. Solid lines represent the total DOS, and the s component is indicated by the shaded area. The arrows show the Fermi energy E_F. (R) and (NR) denote relativistic and nonrelativistic effects, respectively.

On the other hand, reconstruction of surfaces is a much more readily observable effect, involving larger displacements of the surface atoms. It occurs with metal surfaces [13–17] and alloy ones [18]. Unlike relaxation, the phenomenon of reconstruction involves a change in the periodicity of the surface structure which corresponds to an unreconstructed termination of the bulk structure [19,20]. Hence, the influence of the surface reconstruction on the epitaxial layers of metals is possible for gold (5-d metal) because its reconstruction energy is lower than that of copper (3-d metal) as shown in Fig. 3.

The difference in behavior between these two metals suggests the importance of the d bonding of the crystal especially the 5d electronic wave functions which are more localized than the 3d wave functions. The contribution of the d-electrons to the bonding comes from two sources. One is the *d–d* bonding to overlap of *d*-electrons on different atoms on the surfaces deviated from the lattice sites. Another one is the hybridization of the *d*-electrons with the *sp*-electrons, which leads to a lowering of the energy of the occupied d states. Therefore, these lead to a bigger contraction of the lattice and causing a bigger compression of the *sp*-electrons in the bulk.

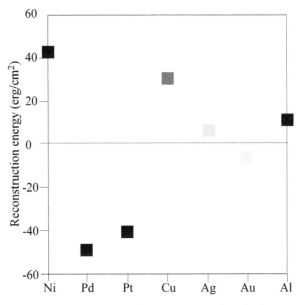

Figure 3 Reconstruction energies for various metals. The larger tendency toward reconstruction of the gold surface is clearly seen than copper.

Summary

Surface reconstruction and relativistic effect may lead to the initial state of the excited electrons of gold and copper. These free electrons may therefore be optically excited by the external field and contribute to the second harmonic generation (SHG). Namely, the SH response at the uppermost atomic layers of the surface of nanowires is dominated by resonant excitation of the collective oscillations of electrons and could be an origin of the SH intensity peak at $2\hbar\omega$ = 3.3 eV for gold nanowires and at 4.4 eV for copper nanowires.

References

[1] K. Locharoenrat, H. Sano, and G. Mizutani, Second harmonic spectroscopy of copper nanowire arrays of on the (110) faceted faces of NaCl crystals, *J. Phys.: Conf. Ser.*, 100 (2008), pp. 052050–052053.

[2] T. Kitahara, A. Sugawara, H. Sano, and G. Mizutani, Optical second-harmonic spectroscopy of Au nanowires, *J. Appl. Phys.*, 95 (2004), pp. 5002–5005.

[3] N. F. Mott and H. Jones, *The Theory of the Properties of Metals and Alloys*, Dover Press, New York, 2000.

[4] P. Pyykkö, Theoretical chemistry of gold, *Angew. Chem. Int. Ed.*, 43 (2004), pp. 4412–4456.

[5] N. E. Christensen and D. L. Novikov, Electronic structure of materials under pressure, *Int. J. Quant. Chem.*, 77 (2000), pp. 880–894.

[6] P. Romaniello and P. L. de Boeij, The role of relativity in the optical response of gold within the time-dependent current-density-functional theory, *J. Phys. Chem.*, 122 (2005), pp. 164303–164310.

[7] J. Autschbach, Perspective: Relativistic effects, *J. Chem. Phys.*, 136 (2012), pp. 150902–150915.

[8] R. Bjornsson and M. Bühl, Electric field gradients of transition metal complexes: Basis set uncontraction and scalar relativistic effects, *Chem. Phys. Lett.*, 559 (2013), pp. 112–116.

[9] H. Häkkinen and M. Moseler, Atom clusters of silver and gold: Symmetry breaking by relativistic effects, *Comp. Mat. Sci.*, 35 (2006), pp. 332–336.

[10] J. David, P. Fuentealba, and A. Restrepo, Relativistic effect on the hexafluorides of group 10 metals, *Chem. Phys. Lett.*, 457 (2008), pp. 42–44.

[11] N. E. Christensen, Relativistic solid state theory, *Theor. Comp. Chem.*, 11 (2002), pp. 863–918.

[12] H. Häkkinen, M. Moseler, and U. Landman, Bonding in Cu, Ag, and Au clusters: Trends and surprises, *Phys. Rev. Lett.*, 89 (2002), pp. 33401–334411.

[13] S. Olivier, G. Tréglia, A. Saúl, and F. Willaime, Influence of surface stress in the missing row reconstruction of fcc transition metals, *Surf. Sci.*, 600 (2006), pp. 5131–5135.

[14] S. Olivier, A. Saúl, and G. Tréglia, Relation between surface stress and (1×2) reconstruction for (110) fcc transition metal surfaces, *Appl. Surf. Sci.*, 212–213 (2003), pp. 866–871.

[15] R. Kempers, P. Ahern, A. J. Robinson, and A. M. Lyons, Modelling the compressive deformation of metal micro-textured thermal interface materials using SEM geometry reconstruction, *Comp. Struct.*, 92–93 (2012), pp. 216–228.

[16] D. P. Woodruff, The role of reconstruction in self-assembly of alkylthiolate monolayers on coinage metal surfaces, *Appl. Surf. Sci.*, 254 (2007), pp. 76–81.

[17] M. Haftel and M. Rosen, Surface-embedded-atom model of the potential-induced lifting of the reconstruction of Au(100), *Surf. Sci.*, 523 (2003), pp. 118–124.

[18] K. L. Man, Y. J. Feng, C. T. Chan, and M. S. Altman, Vibrational entropy-driven dealloying of Mo(100) and W(100) surface alloys, *Surf. Sci.*, 601 (2007), pp. 95–101.

[19] V. Mäkinen and H. Häkkinen, Density functional theory molecular dynamics study of the Au cluster, *Eur. Phys. J.*, 66 (2012), pp. 310–316.

[20] C. J. Heard and R. L. Johnston, Density functional global optimisation study of neutral 8-atom Cu-Ag and Cu-Au clusters, *Eur. Phys. J.*, 67 (2013), pp. 34–40.

Review Paper (2)

Abstract

As nanostructures with well-controlled dimension, composition, and crystallinity are expected to be a new class of intriguing system for investigating structure–property relations, this review article provides a comprehensive review of researches of these materials and related applications.

Introduction

Technology of making small structures, nanofabrication, has evolved greatly over the past decade from a reliance upon clever tricks appropriate for simple structures to a broadly based set of technologies applicable to make complex devices with dimension in the range of one to hundreds of nanometers. Variety of techniques have been used to produce this kind of nanostructures. In this article, the author introduces recent advances in nanofabrication techniques in order to harvest the final nanostructures. The related applications of these nanomaterials are also presented in this review article.

Nanofabrication

Relief structures present on the surface of a solid substrate serve as a class of natural templates for generating supported nanostructures. In this regard, decoration of these templates provides a powerful route to the formation of nanowires made of various metals and semiconductors [1–7]. It can be subsequently transferred onto the surfaces of other substrates. Sugawara et al. have fabricated the arrays of Fe nanowires on the (110) surfaces of NaCl crystals by a shadow deposition method (Fig. 1) [7]. When the NaCl(110) template is annealed in a vacuum condition, the surface becomes faceted with (010) and (100) planes in order to minimize the surface energy. Periodic macrosteps parallel to the [001] direction are then formed. After iron is deposited at a certain degree from the template normal, approximately sixty percent of (100) terraces are exposed to iron flux, and the iron nanowires are formed only near the edge of the ridges.

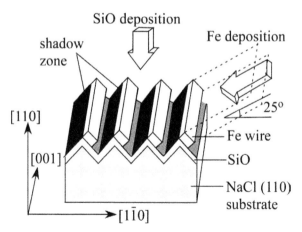

Figure 1 Schematic illustration of a shadow deposition method.

By using the same procedure, they have fabricated cobalt dot arrays sandwiched in gold nanowires onto stepped NaCl(110) surfaces [8]. As demonstrated by Kitahara et al., it is also possible to fabricate arrays of gold nanowires onto the surface of NaCl(110) substrates [9]. The author also uses this simple technique to prepare copper nanowires [10]. Briefly, NaCl(110) single crystals are dipped in distilled water before being mounted on the substrate heating stage in an ultrahigh vacuum deposition chamber. Homoepitaxial layers of NaCl of certain thickness are deposited after the substrate annealing. SiO layers of certain thickness are then deposited as passivation layers. Copper is deposited at a certain deposition rate onto the substrate at a certain flux angle with respect to the surface normal to form copper nanowires. On the arrays of nanowires, SiO layers of certain thickness are deposited as protection layers. For transmission electron microscopy (TEM) observation, the copper nanowire arrays sandwiched by SiO layers are separated from the NaCl substrates by dissolving the substrates in distilled water, and these nanowires are mounted on copper grids for TEM observation. Using this approach, the author obtains wide-area arrays suitable for both TEM and optical measurements in order to study microstructures and optical properties of self-organized arrays of the nanowires.

Penner et al. have demonstrated the growth of the nanowires by templating against the step edges present on a highly oriented pyrolytic graphite (Fig. 2) [11,12]. The noble metal nanowires are found to preferentially nucleate and grow along the step edges present on a graphite surface into a two-dimension parallel array that can be transferred onto the surface of a cyanoacrylate film supported on a glass slide.

The advantage of this technique is that it is applicable to the mass production of high quality metallic nanostructures and applications requiring large areas without breaking vacuum. One drawback of this technique is that the nanowires cannot be peeled off from the substrate to make freestanding form.

Electron beam lithography (EBL) process in Fig. 3 allows one to deposit metal nanowires of any shape, size, and orientation in any arrangement patterns on the flat substrate as described by Schider et al. [13]. In this process, a master image (resist) is produced onto the thin layer of metals, and it is used in transferring a pattern from a mask to surface of the silicon wafer. Finally, the resist is removed by etching in liquid or gaseous form.

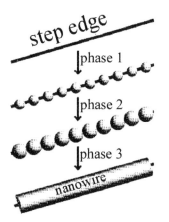

Figure 2 Three phases of nanowires electrodeposition at step edges [10]. 1. Metal nuclei form along step edges. 2. Nucleation ceases; hemispherical nuclei grow to coalescence with nearest neighbors forming rough or beaded nanowires. 3. Beaded nanowires grow.

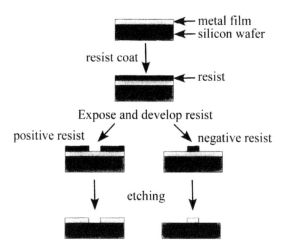

Figure 3 Schematic of main steps of electron beam lithography process.

EBL shows certain benefits over conventional photolithography techniques. It is capable of very high resolution, almost to the atomic level. Typically, EBL has a three orders of magnitude better resolution, although this is limited by the forward scattering of electrons in the resist layer, and back scattering from the underlying substrate. It is a flexible technique that can function with a wide variety of materials and an almost infinite number of patterns. On the other hand, EBL has certain drawbacks. It is slow in operation, being one or more orders of magnitude slower than optical lithography. In addition, it is expensive and complicated, with EBL systems costing many millions of dollars to purchase and require frequent servicing to maintain performance.

Channel in porous membranes provides another class of template for using in the synthesis of nanostructures (Fig. 4). The nanopores in the template are formed by anodizing aluminum films in an acidic electrolyte. Individual nanopore in the alumina is ordered into close-packed honeycomb structure. The diameter of each pore and the separation between two adjacent pores can be controlled by changing the anodizing conditions. Using this membrane template, nanowires of various types of metals and semiconductors can be fabricated. These nanostructures can be deposited into the pores to form continuous nanowires with large aspect ratio (length-to-diameter ratio) by either electrochemical deposition or other methods such as chemical vapor deposition. When freely standing nanowires are desired, one has to remove the template hosts after forming the nanowires in the templates. This task is usually accomplished by dissolving away the template materials in suitable solvent. Sandrock et al. have exploited this membrane-based template to grow gold nanowires with a well-defined dimension [14].

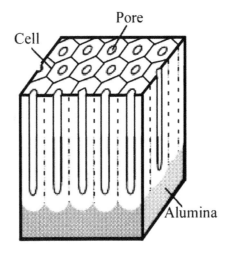

Figure 4 Schematic drawing of an anodic porous alumina template used to form the nanowires by filling the pores with the desired materials.

One advantage of this technique is the possibility of fabricating multilayered structures within nanowires. By varying the cathodic potentials in the electrolyte, layers of different compositions can be controllably deposited. This method provides a low-cost approach to prepare multilayered 1-D nanostructures. One disadvantage of this approach for large-scale applications is that anodic porous alumina is a brittle ceramic film grown on a soft aluminum metal substrate. Great care needs to be exercised in the preparation of the aluminum substrate and in the manipulation of the anodic film to produce pure defect-free porous alumina films that are required to achieve uniform filling of the pores with the nanowires.

Mechanical methods can fabricate the arrays of the atomic-scale metal wires supported on solid substrates by mechanically separating two electrodes in contact. During the separation process, a metal neck is formed between the electrodes due to strong metallic cohesive energy in which it is stretched into an atomically thin wire before breaking. One such method is based on a scanning tunneling microscopy (STM) in which the STM tip is driven into the substrates and the conductance is recorded while the tip is gradually pulled out of the contact with the substrates.

Kawai's Group has employed Au (111), Au (455), and Au (788) crystals as substrates for the growth of the periodic arrays of manganese nanostructures [15]. The gold substrates are chemically etched and mechanically polished to obtain good metal surfaces. Once introduced into an ultrahigh vacuum chamber, the samples are prepared by the certain cycles of sputtering and annealing processes. After cooling the samples to room temperature, manganese is dosed with an electron beam evaporator. A typical deposition rate was about 0.05 milliliters per minute. The coverage of 1 milliliter is defined as one surface layer of manganese crystal.

The advantage of this technique is that one can fabricate a single or an array of stable nanowires. The method can be automated such that a nanowire with a preset quantized conductance can be produced at will. The nanowires supported on a solid substrate can be removed from the fabrication setup and used as a stand-alone device or for further investigation using various experimental probes. However, the use of STM in the setup makes it difficult for mass production. The lifetime of the nanowire is typically less than a few seconds because the gap between the STM tip and the metal surface drifts due to thermal expansion, acoustic noise, and mechanical vibrations, which is also undesirable for practical applications.

Application

The nanostructures have received steadily growing interests as a result of their peculiar and fascinating properties, as well as applications superior to their bulk counterparts [16]. The ability to generate such nanostructures is essential to modern science and technology when making new types of nanostructures, or downsizing the existing microstructures into 1–100 nanometers regimes. The most successful example is provided by microelectronics, where "smaller" has meant greater performance ever since the invention of integrated circuits: more components per chip, faster operation, lower cost, and less power consumption [17]. Miniaturization represents the trend in a range of other technologies. In information storage, for example, there are many active efforts to develop magnetic and optical storage components with critical dimensions as small as tens of nanometers [18]. It is also clear that a new phenomenon is associated with nanometer-sized structures, with the best-established examples including size-dependent excitation of the quantum dot structures [19], and quantization of conductance in the metal contacts [20]. In addition, quantum confinement of electrons by the potential wells of nanoscale structures may provide one of the most powerful means to control the electrical, optical, magnetic, and thermoelectric properties of a solid state functional material.

So far, nanostructures such as wires, rods, belts, and tubes have become the focus of intensive research owing to their fabrication of nanodevices [21]. These nanostructures provide a good system to investigate the dependence of electrical and thermal transport or mechanical properties on the size reduction. They are also expected to play an important role in the interconnect applications and the functional units in fabricating electronic, optoelectronic, electrochemical, and electromechanical devices.

Although nanostructures can now be produced using a number of advanced nanolithography techniques [22,23], such as electron beam, these methods are still not conveniently available to practical routes, and it is also challenging to fabricate nanowires with the widths of a few nanometer. Exploration and development of new techniques for fabrication of large quantities of nanostructures from a diversified range of materials, rapidly, and at reasonably low costs, are desirable.

Many deposition techniques described herein are not intended to replace the existing methods, but rather to provide a simple alternative for generating nanostructures with a well-controlled dimension, and to help the study of their optical properties.

Conclusion

Recent advances in nanofabrication techniques have made it possible to produce well-defined nanostructures and each method has its specific merits and inevitable weakness. For instance, some researchers devote the most attention to the method of generating nanostructures based on anisotropic growth [24] directed or confined by the templates. The template-directed methods provide a good control over the uniformity and dimension in which the nominal thicknesses of the nanowires are measured with the sensor plane of the thickness monitor perpendicular to the metal beam direction. However, removal of the template through a post-synthesis process may cause damage to the final nanostructures' product. Most nanostructures produced using this method are polycrystalline in structure and they may limit their uses in device fabrication and fundamental studies. Therefore, judging against these aspects, some methods described here still need to be improved before it finds widespread use in commercial applications.

References

[1] Hu L, Wu H, and Cui Y. 2011. *MRS Buletin*, 36, pp. 760.

[2] Neubrech F, Kolb T, Lovrincic R, Fahsold G, and Pucci. 2006. *Appl. Phys. Lett.*, 89, pp. 253104.

[3] He H and Tao N J. 2003. *Encyclopedia Nanosci. Nanotech.*, 4, pp. 1.

[4] Wu Y, Xiang J, Yang C, Lu W, and Lieber A. 2004. *Nature*, 430, pp. 61.

[5] Tian M, Wang J, Kurta J, Mallouk T E, and Chan M H W. 2003. *Nano Lett.*, 3, pp. 919.

[6] Vossen J L and Kern W. 1978. *Thin Film Processes* (New York: Academic).

[7] Sugawara A, Coyle T, Hembree G G, and Scheinfein M R. 1997. *Appl. Phys. Lett.*, 70, pp. 1043.

[8] Kitahara T, Sugawara A, Sano H, and Mizutani G J. 2004. *J. Appl. Phys.*, 95, pp. 5002.

[9] Kitahara T, Sugawara A, Sano H, and Mizutani G. 2003. *Appl. Surf. Sci.*, 219, pp. 271.

[10] Locharoenrat K, Sugawara A, Takase S, Sano H, and Mizutani G. 2007. *Surf. Sci.*, 601, pp. 4449.

[11] Penner R M. 2002. *J. Phys. Chem. B*, 106, pp. 3339.

[12] Song H H, Jones K M, and Baski A A. 1999. *J. Vacuu. Sci. Technol. A*, 17, pp. 1696.

[13] Schider G, Krenn J R, Gotschy W, Lamprecht B, Ditbacher H, Leitner A, and Aussenegg F R. 2001. *J. Appl. Phys.*, 90, pp. 3825.

[14] Sandrock M L, Pibel C D, Geiger F M, and Foss Jr C A. 1999. *J. Phys. Chem. B*, 1032668.

[15] Shiraki S, Fujisawa H, Nantoh M, and Kawai M. 2004. *Surf. Sci.*, 552, pp. 243.

[16] Nalwa H S. 2000. *Handbook of Nanostructured Materials and Nanotechnology* (New York: Academic).

[17] Luryi S, Xu J, and Zaslavsky A. 1999. *Future Trends in Microelectronics* (New York: John Wiley & Sons).

[18] Ross C. 2001. *Annu. Rev. Mat. Sci.*, 31, pp. 203.

[19] Alivisatos A P. 1996. *Science*, 271, pp. 933.

[20] Krans J M, Rutenbeek J M, Fisun V V, Yanson I K, and Jongh L J. 1995. *Nature*, 375, pp. 767.

[21] Wang Z L. 2000. *Adv. Mat.*, 12, pp. 1295.

[22] Krenn J R, Weeber J C, Dereux A, Bourillot E, Goudonnet J P, Schider B, Leitner A, Aussenegg F R, and Girard C. 1999. *Phys. Rev. B*, 60, pp. 5029.

[23] Xia Y, Rogers J A, Paul K E, and Whitesides G M. 1999. *Chem. Rev.*, 99, pp. 1823.

[24] Growth anisotropy describes the condition when growth rates are not equal in all directions. In contrast, when growth rates are the same rate in all directions, growth is isotropic.

Review Paper (3)

Abstract

Over the past few years significant progress has been made in investigations into the synthesis of nanowires of various materials and their applications to nanoscale electronics, optical devices, and sensing systems; nevertheless, a major hurdle facing the developers of the integrated nanowire systems is organizing, manipulating, or controlling the placement of large numbers of nanowires. These problems can be solved by using a shadow deposition method for fabrication of polycrystalline metallic nanowires. This technique uses topographical features of ridge-to-valley to expose specific areas of the alkali halide substrates in order to grow the nanowire arrays. Nanowires of large areas show a preference for formation along exposed steps in the atomic structure of the substrate. The minimum widths of the obtained nanowires are less than 40 nm.

The strongly geometrical shape, high aspect ratios, and nanoscale cross-section of metallic nanowires can be expected to affect optical properties through confinement effects. Optical properties of the obtained metallic nanowires are then investigated by studies of linear reflectivity spectra, linear absorption spectra, and second harmonic generation spectra. Theoretical calculations by Maxwell–Garnett model and finite-difference time-domain method are also represented for predicting the reflectivity, absorption, and the electric field distributions close to the wires. With experimental and theoretical studies, it is discovered that there are differences of optical properties among these metallic nanowires. These will be one of the most important issues when considering the types of materials used in current applications and development of new applications.

Introduction

There are many potential applications for which metallic nanowires may become important: for electronic nanodevices and for metallic interconnects of quantum nanodevices. These applications in any devices require the large areas with high throughput of the metallic nanowires. A shadow deposition method is considered as one of the promising procedures to fulfill this requirement. In contrast to a lithographic method [1], this self-organized pattern formation is also completed without breaking vacuum [2].

The primary role of the small size of the metallic nanowires shows a wealth of optical phenomena. The external light photons can couple with the electrostatic field of plasmons generating the strong absorption band [3]. With the development of pulsed laser technology, nonlinear optical techniques including second harmonic generation (SHG) are also used to understand the anisotropic nature of these nanomaterials. A great advantage of these studies over the linear optical techniques is its intrinsic surface sensitivity due to the fact that the SH signal originates from a thin interfacial region [4–6]. Owing to this surface sensitivity, the resonant coupling of surface plasmons in the metallic nanowires with SHG may result in an enhancement of the electromagnetic field. The linear dielectric response of the composites of the metallic nanowires in the dielectric medium may lead to an enhancement of second order optical nonlinearity. These factors motivate the need for further studies toward the elucidation of the microscopic origin of the nonlinearity of metallic nanostructures.

In this chapter, three types of metallic nanowires are fabricated by a shadow deposition of metal atoms on a faceted NaCl(110) template. The microstructures of the nanowires are analyzed by means of transmission electron microscopy (TEM). The linear and nonlinear optical properties of the arrays of the nanowires are then investigated. Finally, the comparison between the experiments and the theories by Maxwell–Garnett model and finite-difference time-domain (FDTD) method is discussed.

Experiment

NaCl(110) single crystals having the dimension of $10 \times 18 \times 4$ mm^3 are dipped in distilled water for 10 seconds and in ethanol for 1 minute in two cycles before being mounted on the substrate heating stage in an ultrahigh vacuum deposition chamber with a base pressure of approximately 2×10^{-9} Torr. The 200–300 nm thick homoepitaxial layers of NaCl are deposited after the substrate annealing at 200–450°C. The 5–10 nm thick SiO layers are then deposited as passivation layers. Desired metal is deposited at the given deposition rate onto the substrate at the flux angle of 65° with respect to the surface normal to form nanowires. On the arrays of metallic nanowires, 5–10 nm thick SiO layers are deposited as protection layers. For transmission electron microscopy (TEM) observation, metallic nanowire arrays sandwiched by SiO layers are separated from the NaCl substrates by dissolving the substrates in distilled water, and mounted on 200–400 mesh Cu grids. Fig. 1 gives a brief outline of the shadow deposition procedure.

Nanowire Characterization. Transmission electron microscopy (TEM) images in Fig. 2(a), (b), and (c) show the uniform Au, Cu, and Pt nanowires, respectively, as long and straight periodic dark images parallel to the [001] direction of the substrates. The nominal thicknesses of the nanowires are measured with the sensor plane of the thickness monitor set perpendicular to the metal beam direction and showed in Table 1. The minimum widths of the obtained nanowires are 40 nm for Au, 14 nm for Cu, and 9 nm for Pt.

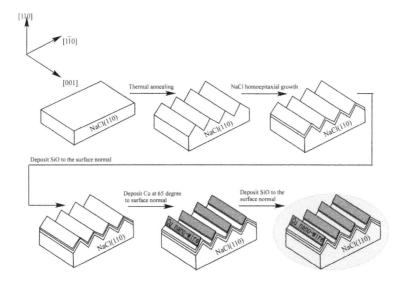

Figure 1 Schematic view of a shadow deposition method.

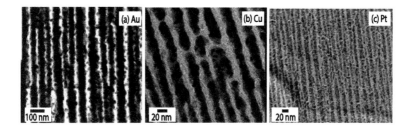

Figure 2 Transmission electron microscopy (TEM) images of Au (a), Cu (b), and Pt (c) nanowire arrays after the NaCl template has been dissolved. The metallic nanowire arrays are seen in dark contrast running parallel to the [001] direction of the NaCl (110) template. The minimum widths of the obtained nanowire arrays are 40 nm for Au, 14 nm for Cu, and 9 nm for Pt.

One of the key parameters determining the minimum width of the nanowires deposited onto a crystalline substrate is the surface energy of the wire metal (Fig. 3). High surface energy of metals leads to high contact angle on the substrate surface as shown in Fig. 4 [7]. As the surface energy is the largest for Pt, the second largest for Cu, and the smallest for Au in their liquid state, the minimum width of Cu nanowires (14 nm) is larger than that of Pt (9 nm), but is smaller than that of Au (40 nm).

Table 1 Geometrical parameters of the fabricated metallic nanostructures

Nanowires	Nominal thickness (nm)	Wire width (nm)	Periodicity (nm)
Au	65	40	80
Cu	10	14	27
Pt	7	9	17

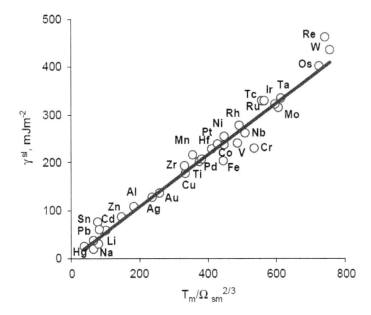

Figure 3 Surface energy of metals at melting point γ^{sl} with respect to $T_m/\Omega_{sm}{}^{0.7}$. T_m and Ω_{sm} are melting point and atomic volume of solid at melting point, respectively.

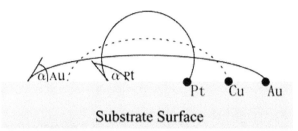

Substate Surface

Figure 4 Equilibrium shape of a spherical drop from different contact angles of metals on flat solid substrate (α).

Optical Properties. Reflectivity and absorption spectra can be measured as a function of the sample rotation angle. These spectra are recorded in the wavelength 250–1100 nm by a spectrometer equipped with a polarizer accessory, a Xenon lamp source, and a photomultiplier (Fig. 5).

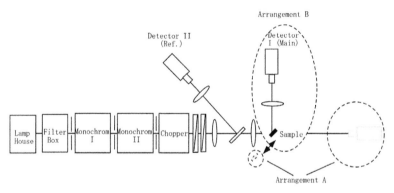

Figure 5 Schematic diagram of linear optical experimental setup. Arrangements A and B are applied to measure absorption and reflectivity spectra, respectively.

A drop in linear reflectivity spectra of Au nanowires is observed around $\hbar\omega = 1.7$ eV for the incident field perpendicular to the wire axes in Fig. 6(a), but it is not observed for incident field parallel to the wire axes. In Fig. 6(c), the drop in the reflectivity spectra near 4.2 eV for Pt nanowires is observed for the electric field polarized perpendicular to the wire axes and they shift to lower photon energy for the parallel polarization.

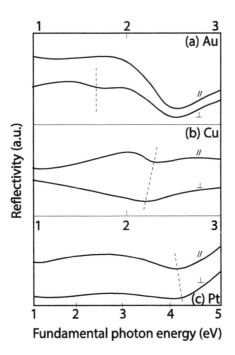

Figure 6 Measured linear reflectivity spectra of Au (a), Cu (b), and Pt (c) nanowire arrays. $\perp$ (//) means the electric field perpendicular (parallel) to the wire axes.

Cu nanowires exhibit reflectivity-polarization angle trends opposite to those of Pt nanowires. The drop in reflectivity near 2.2 eV is observed for the electric field polarized perpendicular to the wire axes in Fig. 6(b) and they shift to higher photon energy for the parallel polarization. Interestingly, the drop in reflectivity spectra near 2.2 eV for Cu nanowires in Fig. 6(b) appears to have a similar origin as the main absorption peak at 2.3 eV for Cu nanowires in Fig. 7. Absorption maxima are remarkably observed near 2.3 eV for the perpendicular polarization in Fig. 7 and they shift to lower energy for the parallel polarization. This possibly may be a plasmon excitation in the nanostructures giving rise to the local electric field enhancement.

Theoretical analysis of linear reflectivity and absorption is performed in order to find whether the plasmon resonances served as the origin of the enhancement of the electromagnetic field from these metallic nanowires. The reflectivity and absorption of nanowires are calculated according to Maxwell–Garnett model [8–10]. For Cu nanowires, for instance, it is found that the detailed experimental dependence of the energy positions of the plasmon maxima on the polarization is different from the prediction by Maxwell–Garnett model [3]. When the electric field is perpendicular to the wire axes, the calculated positions of plasmon maxima are observed near 2.0 eV in contrast to the experiment showing plasmon maxima around 2.3 eV as seen in Fig. 7. When the electric field is parallel to the wire axes, plasmon maximum energies are observed below 2.0 eV, and they conflict with the experiment in Fig. 7 showing plasmon maxima above 2.0 eV. Another rigorous numerical study using the finite-difference time-domain (FDTD) method of solving the electromagnetic problems by integrating Maxwell's differential equations is then introduced [11,12].

The FDTD method allows one to consider the actual wire size dependence of the peak energy positions for metallic nanowires [3]. This simulation is more favorable than that of Maxwell–Garnett model from a point of view of the reproduction of the experimental data. The calculation results show the plasmon maxima near 2.4 eV for Cu nanowires as the electric field is perpendicular to the wire axes, while the blue-shift occurs when the electric field is paralleled to the wire axes. The peak positions in the latter polarization configuration are above 2.0 eV, as is consistent with the experimental results in Fig. 7. This present calculation taking into account the absolute wire size by FDTD simulation is judged to have improved the agreement between the experiment and the theory because it takes full account of the retardation effect.

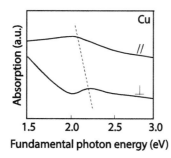

Figure 7 Measured linear absorption spectra of Cu nanowire arrays. ⊥ (//) means the electric field perpendicular (parallel) to the wire axes.

From the discussion so far, the theoretical results indicate that the resonant enhancement near the fundamental photon energy of 1.7 eV for Au, 2.2 eV for Cu, and 4.2 eV for Pt possibly originates from the coupling of plasmon resonances in the metallic nanowires; however, these findings will be confirmed by their nonlinear optical properties in the last section of this chapter.

Angle Dependence. The probe-light pulses with a fundamental photon energy of 1.17 eV for Au and Cu (2.23 eV for Pt) can be generated by a mode-locked Nd:YAG laser (Fig. 8). The fundamental light beam is focused onto the sample surface and the incident angle is 45°. The second harmonic (SH) signal is detected by a photomultiplier and the absolute SH intensities are obtained by normalizing the SH response to that of the α-SiO$_2$ (0001). To measure the azimuthal angle dependence of the SH intensity, the sample is mounted on an automatic rotation stage with the surface normal set parallel to the rotation axis of the stage. The measurements are carried out in four different input/output polarization combinations: p-in/p-out, p-in/s-out, s-in/p-out, and s-in/s-out. All SHG observations are carried out in air at room temperature.

Azimuthal angle dependence of the SH intensity from the Au, Cu, and Pt nanowire arrays is measured and showed in Fig. 9. The SH intensity is measured as a function of the sample rotation angle ϕ at the photon energy of 1.17 eV for Au and Cu and at the photon energy of 2.33 eV for Pt. The fundamental output at 1.17 eV cannot be used for Pt nanowires because they are easily damaged due to their strong optical absorption. The angle ϕ is defined as the angle between the incident plane and the wire axes in the [001] direction. In each row, the SH intensity patterns for four different configurations of input and output polarizations are showed for each sample.

The SH intensity from metallic nanowires depends strongly on the sample rotation angle ϕ and the polarization configurations of the fundamental and SH light. The patterns of the SH intensity from Au nanowires show two lobes for the p-in/p-out configuration in Fig. 9(a), four lobes for the p-in/s-out polarization configuration in Fig. 9(b), and two lobes for the s-in/p-out polarization configuration in Fig. 9(c). The

SH intensity patterns from Cu nanowires display two main lobes for the p-in/p-out, s-in/p-out, and s-in/s-out polarization configurations in Fig. 9(e), (g), and (h), respectively. For Pt nanowires, two lobes are seen in any incident and output light polarizations as shown in Fig. 9(i)–(l).

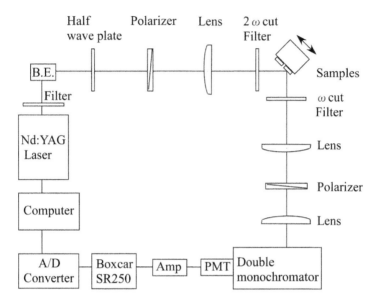

Figure 8 Block diagram of the experimental setup for measuring SHG rotational anisotropy. The fundamental laser beam is incident on the sample at an angle of 45°, and SH photons are collected in reflection as a function of the sample rotation angle ϕ. The ϕ is defined as the angle between the plane of incidence and the [001] direction of the sample surface. When $\phi = 270°$ the incident plane is defined in the same direction as the desired metal flux during the deposition. The rotational anisotropy in SH intensity patterns is obtained with different input and output polarization configurations. The B.E., PMT, Amp, and A/D stand for beam elevator, a photomultiplier, amplifier, and analogue-to-digital converter, respectively.

The SH intensity patterns from these metallic nanowires in Fig. 9 (filled circles) are analyzed by a phenomenological model [13,14]. The fitted results are showed in the solid curves in Fig. 9. The rotation-angle dependences of the mainly second-order susceptibility tensor elements for different metallic nanowires are then indicated in Table 2.

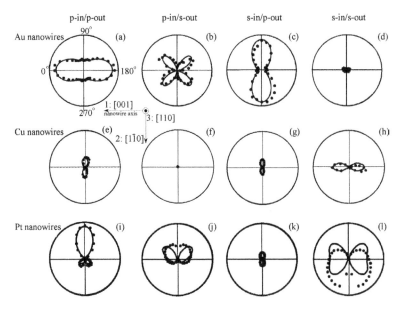

Figure 9 Measured (filled circles) and calculated (solid line) second harmonic (SH) intensity patterns for different metallic nanowire arrays as a function of the sample rotation angle ϕ. The fundamental photon energy is 1.17 eV for Au and Cu nanowire arrays, 2.23 eV for Pt nanowire arrays. The angle ϕ is defined as the angle between the incident plane and the [001] direction on the sample surface. The incident angle is 45°. The SH intensity is plotted in polar coordinates. The four different input and output polarization configurations are indicated at the top. The intensity scales are the same for the four polarization configurations for each sample.

Table 2 Azimuthal angle dependences of the mainly second-order susceptibility tensor elements for different metallic nanowire arrays

Nanowires	p-in/p-out	p-in/s-out	s-in/p-out	s-in/s-out
Au	$\chi^{(2)}_{113}, \chi^{(2)}_{223}, \chi^{(2)}_{333}$	$\chi^{(2)}_{113}, \chi^{(2)}_{223}$	$\chi^{(2)}_{311}, \chi^{(2)}_{333}$	
Cu	$\chi^{(2)}_{323}$		$\chi^{(2)}_{311}$	$\chi^{(2)}_{222}$
Pt	$\chi^{(2)}_{113}, \chi^{(2)}_{223}, \chi^{(2)}_{323}$	$\chi^{(2)}_{113}, \chi^{(2)}_{223}$	$\chi^{(2)}_{311}$	$\chi^{(2)}_{222}$

It is found that the main nonlinear susceptibility elements $\chi^{(2)}_{113}$, $\chi^{(2)}_{223}$, $\chi^{(2)}_{311}$, and $\chi^{(2)}_{333}$ dominate the SH signal for Au nanowires. The indices 1, 2, and 3 denote the [001], [1$\bar{1}$0], and [110] directions on the NaCl(110) surface, respectively. The major contributions of the $\chi^{(2)}_{323}$, $\chi^{(2)}_{311}$, and $\chi^{(2)}_{222}$ elements are proved to be prominent in the SH signal for Cu nanowires, whereas the main $\chi^{(2)}_{113}$, $\chi^{(2)}_{223}$, $\chi^{(2)}_{222}$, $\chi^{(2)}_{311}$, and $\chi^{(2)}_{323}$ elements are dominant in the SH signal for Pt nanowires. The positive and negative directions of the 2: [1$\bar{1}$0] direction for Pt nanowires in Fig. 9(i) are not equivalent due to the geometrical nature of the shadow deposition.

The $\chi^{(2)}_{113}$ and $\chi^{(2)}_{223}$ elements dominate second order optical nonlinearity of Au nanowires for the p-in/s-out polarization configuration in Fig. 9(b) partly due to a strong depolarization field in the wires. This similar effect is observed for Pt nanowires in Fig. 9(j), but it is not for Cu nanowires in Fig. 9(f). The prominent $\chi^{(2)}_{113}$ and $\chi^{(2)}_{223}$ elements also lead to an enhancement of the optical nonlinearity of Au and Pt nanowires for the p-in/p-out polarization configuration. As clearly seen in this polarization configuration, the depolarization effect from Au nanowires in Fig. 9(a) is judged to be stronger than that from Pt nanowires in Fig. 9(i). The depolarization effect has rarely been observed as the origins of optical enhancement for Cu nanowires in Fig. 9(e). The two-lobed patterns in Au nanowires are wider in the middle in p-in/p-out (Fig. 9(a)) and s-in/p-out (Fig. 9(c)) polarization configurations due to the contribution from the $\chi^{(2)}_{333}$ element. Suppression of the $\chi^{(2)}_{113}$ and $\chi^{(2)}_{223}$ elements for Cu nanowires is existed due to a weak depolarization field in the wires leading to the intensity of SHG as low as the noise level for the p-in/s-out polarization configuration in Fig. 9(f).

The pattern for Cu nanowires for the s-in/s-out polarization configuration in Fig. 9(h) is seen to be dominated by the contribution of the $\chi^{(2)}_{222}$ element. This element results from the broken symmetry in the 2: [1$\bar{1}$0] direction due to the cross-sectional shapes of the nanowires produced by the periodic macrosteps parallel to the 1:[001] direction. The similar effect is observed for Pt nanowires in Fig. 9(l). However, it is not observed for Au nanowires in Fig. 9(d) because the incident electric field in the 2:[1$\bar{1}$0] direction is partly canceled by the depolarization field perpendicular to the wire axes. This difference is partly due to the larger anisotropy in the nonlinearity of Cu and Pt nanowires due to thinner wires.

Local electromagnetic field is numerically investigated by FDTD method in order to clarify the contributions of the $\chi_{311}^{(2)}$ and $\chi_{323}^{(2)}$ elements dominated the optical nonlinearity of the nanowires [3]. It is found that the contribution of the $\chi_{323}^{(2)}$ element originating from local plasmons in Cu nanowires dominates the optical nonlinearity for the p-in/p-out polarization configuration in Fig. 9(e), and the contribution of the $\chi_{311}^{(2)}$ element originating from the lightning-rod effect dominates the optical nonlinearity for the s-in/p-out polarization configuration in Fig. 9(g). These similar effects also lead to an enhancement of the optical nonlinearity of Pt nanowires. The contribution of the $\chi_{323}^{(2)}$ element in Pt nanowires originating from the plasmon resonances dominates the optical nonlinearity for the p-in/p-out polarization configuration in Fig. 9(i), and the contribution of the $\chi_{311}^{(2)}$ element originating from the lightning-rod effect dominates the optical nonlinearity for the s-in/p-out polarization configuration in Fig. 9(k). For Au nanowires, it is possible that the contribution of the $\chi_{311}^{(2)}$ element originating from the lightning-rod effect dominated the optical nonlinearity for the s-in/p-out polarization configuration in Fig. 9(c). The contribution of the $\chi_{323}^{(2)}$ element originating from the plasmon resonances is not observed for the p-in/p-out polarization configuration in Fig. 9(a) due to a strong depolarization field in Au nanowires.

From the analysis of the SH intensity patterns by a phenomenological model and FDTD method, it can be concluded that SHG from Cu (Pt) nanowires is found to be different from those of Au nanowires. This is due to the weaker suppression of the incident field by the depolarization field in Cu (Pt) nanowires and the stronger anisotropy in the nonlinearity of Cu (Pt) nanowires due to thinner wires than Au. The electric field component along the surface normal enhanced by the plasmon-resonant coupling in the wires is observed for Cu (Pt) nanowires, but it is not easily noticed for Au nanowires.

Wavelength Dependence. Light source of the fundamental photon energy from 1.2 to 2.3 eV can be an optical parametric generator and amplifier system driven by a mode-locked Nd:YAG laser (Fig. 10). The excitation light is focused on the sample surface at the incident angle of 45°. The SH signal from the sample is detected by a photomultiplier. Finally, the absolute magnitudes of the SH intensity are obtained by normalizing the SH response to that of the α-SiO$_2$ (0001). The measurements are performed in p-in/p-out polarization configuration. All SHG observations are performed in air at room temperature.

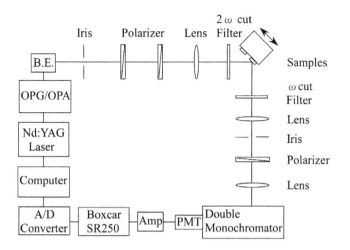

Figure 10 Block diagram for measuring the SH intensity as a function of the photon energy. The wavelength tunable light source is the optical parametric generator and amplifier (OPG/OPA) system and it is incident on the sample at an angle of 45°. The B.E., PMT, Amp, and A/D represent beam elevator, a photomultiplier, amplifier, and analogue-to-digital converter, respectively.

The SH intensity spectra from metallic nanowires are measured as a function of the SH photon energy for the p-in/p-out polarization combination as shown in Fig. 11. The incident wave vector of the excitation is parallel to the $[1\bar{1}0]$ direction of the NaCl(110) faceted substrates. The SH intensity spectra taken from Au nanowires in Fig. 11(a) show a nearly flat dependence of SH intensity on the photon energy below $2\hbar\omega = 3.0$ eV. The intensity abruptly increased above $2\hbar\omega$ ~ 3.1 eV and reached a maximum at 3.3 eV. The SH intensity spectra for Cu nanowires in Fig. 11(b) showed steady increase above $2\hbar\omega = 3.0$ eV. The SH response exhibits a peaked resonance near 4.4 eV. In Fig. 11(c), the SH intensity spectra for Pt nanowires show steady increase above $2\hbar\omega = 3.7$ eV. The SH response exhibits a peaked resonance near 4.3 eV.

The main peak near $2\hbar\omega = 3.3$ eV for Au nanowires in Fig. 11(a) appears to have a similar origin as the main peak at 4.4 eV for Cu nanowires in Fig. 11(b), and the main peak at 4.3 eV for Pt nanowires in Fig. 11(c). Two candidate origins of the SHG from these metallic nanowires are considered. They are (1) the enhancement of the local field by the plasmon excitation in the nanowires and (2) the dielectric response at the metal/SiO interface.

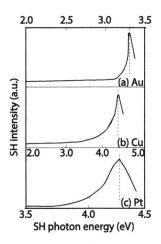

Figure 11 Measured second harmonic (SH) intensity spectra from different metallic nanowire arrays as a function of the SH photon energy for the p-polarized input and p-polarized output configuration. The incident wave vector of excitation is parallel to the [1$\bar{1}$0] direction of the NaCl(110) faceted substrates.

Let us discuss a resonance enhancement of the SH signal with the dipolar mode of the surface plasmon in the metal wires. Theoretical analysis of the SH intensity is performed in order to determine the values of the local field factor and then to find whether the surface plasmon resonance served as the origin of the enhancement of the SH response from the metallic nanowires. Local field enhancement factor for Au, Cu, and Pt nanowires is calculated by a quasi-static theory [9]. The calculated enhancement factor is described in Refs. [3–16]. It is found that the calculated spectra for the metallic nanowires in Fig. 12 reproduce the experimental findings in Fig. 11 well. Hence, these theoretical results indicate that the resonant enhancement near the SH photon energy of 3.3 eV for Au, 4.4 eV for Cu, and 4.3 eV for Pt originate from the coupling of plasmon resonances in the metal nanowires. Experimental findings and calculations are consistent with the suggestion by Schider et al. that the plasmon excitation in metallic nanostructures giving rise to the local electric field enhancement [2]. For Au nanowires, it is noted that the nonlinear response was weak at the SH photon energy lower than 3.3 eV, due to the anisotropic depolarization field.

Next, we consider an enhancement of the optical nonlinearity of metallic nanowires assigned to the electronic resonance at the metal/

SiO interface. For Au nanowires, the SH intensity spectrum has a main peak around $2\hbar\omega$ = 3.3 eV in Fig. 11(a). It is suggested that the Fresnel factor for the incident light creates a structure at 1.7 eV due to the linear dielectric structure seen in the linear reflectivity in Fig. 6(a), and enhances the SH intensity at twice this photon energy. Like Au nanowires, the SH intensity spectrum of Cu nanowires has a main peak around $2\hbar\omega$ = 4.4 eV in Fig. 11(b). It is also suggested that the Fresnel factor for the incident light creates a structure at 2.2 eV due to the linear dielectric structure seen in the linear reflectivity in Fig. 6(b), and enhanced the SH intensity at twice this photon energy. In contrast, the drop in linear reflectivity spectra for Pt nanowires is observed near 4.2 eV in Fig. 6(c). It is possible that a dielectric structure exists near $2\hbar\omega$ = 4.3 eV for Pt nanowires in Fig. 11(c) and the Fresnel factor at frequency 2ω for SHG is enhanced by one photon resonance of SH photons. Hence, these findings clearly suggest that the observed structures in the SH spectra of metallic nanowires are explained as a result of the linear dielectric response of the composites of the metallic nanowires in SiO matrix. They lead to an enhancement of second order optical nonlinearity of metallic nanowires.

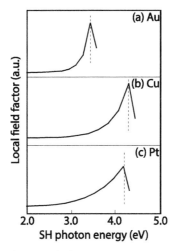

Figure 12 Calculated second harmonic generation (SHG) local field enhancement factor as a function of the SH photon energy for different metallic nanowire arrays according to Ref. [9].

From the analysis of the SH intensity by a quasi-static theory, one can conclude that the SHG from metallic nanowires is originated from the plasmon resonance and the linear dielectric properties of the metal/SiO interface.

Conclusion

Microstructures and optical properties of self-organized arrays of the metallic polycrystalline nanowires, Au, Cu, and Pt, are studied. Shadow deposition onto (110) faceted faces of NaCl crystals is demonstrated as a simple technique for fabricating wide-area arrays of the nanowires suitable for both transmission electron microscopy and optical measurements. It is found that there are two candidate origins of the observed SH intensity from metallic nanowires. One is the enhancement of the local field by the plasmon excitation in the nanowires. Another one is the dielectric structure at the metal/SiO interface. The second harmonic generation (SHG) from Cu (Pt) nanowires is found to be different from those of Au nanowires. This is due to the weaker suppression of the incident field by the depolarization field in Cu (Pt) nanowires and the stronger anisotropy in the nonlinearity of Cu (Pt) nanowires due to thinner wires. This difference may have significant consequences in the range of applications that such materials may be useful for.

References

[1] Sugawara A., Hembree G. G., and Scheinfein M. R. *Journal of Applied Physics*, 1997, 82, pp. 5662–5669.

[2] Schider G., Krenn J. R., Gotschy W., Lamprecht B., Ditlbacher H., Leitner A., and Aussenegg F. R. *Journal of Applied Physics*, 2001, 90, pp. 3825–3830.

[3] Locharoenrat K., Sugawara A., Takase S., Sano H., and Mizutani G. *Surface Science*, 2007, 601, pp. 4449–4453.

[4] Shen Y. R. *The Principles of Nonlinear Optics*; Wiley: New York, 2002, pp. 5–50.

[5] Bloembergen N. *Nonlinear Optics*; World Scientific: Miami, 2005, pp. 25–77.

[6] Boyd R. W. *Nonlinear Optics*; Academic: Orlando, 2008, pp. 55–102.

[7] Digilov R. M. *Physica B*, 2004, 352, pp. 53–60.

[8] Maxwell-Garnett J. C. *Philosophical Transactions* of the Royal Society London A, 1904, 203, pp. 385–420.

[9] Al-Rawashdeh N. A. F., Sandrock M. L., Seugling C. J., and Foss J. C. A. *Journal of Physical Chemistry B*, 1998, 102, pp. 361–371.

[10] Hayashi N., Aratake K., Okushio R., Iwai T., Sugawara A., Sano H., and Mizutani G. *Applied Surface Science*, 2007, 253, pp. 8933–8938.

[11] Sullivan D. M. *Electromagnetic Simulation Using the FDTD Method*; IEEE: New York, 2013, pp. 7–105.

[12] Taflove A. *Computational Electrodynamics: The Finite-Difference Time-Domain Method*; John Wiley & Sons: New Jersey, 2010, pp. 13–108.

[13] Sipe J. E., Moss D. J., and van Driel H. M. *Physical Review B*, 1987, 35, pp. 1129–1141.

[14] Kobayashi E., Mizutani G., and Ushioda S. *Japanese Journal of Applied Physics*, 1997, 36, pp. 7250–7256.

[15] Kitahara T., Sugawara A., Sano H., and Mizutani G. *Journal of Applied Physics*, 2004, 95, pp. 5002–5005.

[16] Takabe H., Luong N. H., and Onuki Y. *Frontiers of Basic Science Towards New Physics, Earth and Space Science, Mathematics*; Osaka University Press: Osaka, 2006, pp. 53–104.

Review Paper (4)

Abstract

Although absorption is still the primary optical property of interest, other spectroscopic techniques including optical second harmonic generation (SHG) are also used to understand anisotropic nature of nanomaterials. Benefit of SHG technique as compared with linear optical approach is that SH response is the most sensitive to surface potential changes and SH intensity mainly comes from surface/interface layers. This article reports that optical absorption spectra due to surface plasmon modes of metallic nanowires exhibit strong anisotropic absorption. Maxwell–Garnett simulation cannot fit with experimental results because of wire sizes much larger than wavelength of light, whereas finite-difference time-domain calculation can fit well with experimental findings in terms of position of photon energy of absorption peak with respect to polarization conditions.

Introduction

In solid state materials, there are electrons localized close to positive ions and free electrons contributing to electrical conductivity. Interaction of electrons with materials is reflected in dielectric function. We can also describe the effect of free electrons in the dielectric function. Lorentz model describes the dielectric properties of insulators most simply, in which electrons are bound to positive ions. Equation of motion of an electron in an oscillating electric field of frequency ω is

$$m\left[\frac{d^2x}{dt^2}+\Gamma_0\frac{dx}{dt}+\omega_0^2x\right]=qE \tag{1}$$

$$m\left[\frac{d^2x}{dt^2}+\Gamma_0\tau\frac{dx}{dt}+\omega_0^2x\right]=qE_0e^{-i\omega t} \tag{2}$$

The m is free electron mass, x is displacement, and Γ_0 is damping constant. The τ is time interval between one scattering and another of free electron, and is normally about 10 fs. The ω_0 is frequency of a transverse wave and q is electric charge. The E is electric filed giving rise to polarization.

When we assume a solution of the form

$$x = x_0e^{-i\omega t} \tag{3}$$

The x_0 is found as

$$x_0 = \frac{qE_0}{m(\omega_0^2-\omega^2-i\omega\Gamma_0)} \tag{4}$$

In metals and semiconductors (Drude model), if we suppose that there is no restoring force applied to free electron, substituting ω_0 is equal to zero. Equation (2) is modified as

$$m*\left[\frac{d^2x}{dt^2}+\tau^{-1}\frac{dx}{dt}\right]=qE_0e^{-i\omega t} \tag{5}$$

The $m*$ is effective mass. Dielectric function of this material of volume fraction V is

$$\varepsilon = \varepsilon_0 - \frac{Ne^2}{m*V\omega(\omega+i\tau^{-1})} \tag{6}$$

The N is electron concentration.

On one hand, when $\omega \sim 0$, Eq. (6) becomes

$$\varepsilon = \varepsilon_0 + i\frac{Ne^2\tau}{m*V\omega} \tag{7}$$

If electrical conductivity is defined as

$$\sigma = \frac{Ne^2\tau}{m*V} \tag{8}$$

Eq. (7) reads

$$\varepsilon = \varepsilon_0 + i\frac{\sigma}{\omega} \tag{9}$$

On the other hand, when $\omega \gg \tau^1$, Eq. (6) becomes

$$\varepsilon = \varepsilon_0 - \frac{Ne^2}{m*V\omega^2} \tag{10}$$

If $\quad \omega_p = \sqrt{\frac{Ne^2}{m*\varepsilon_0 V}} \tag{11}$

one reads $\quad \varepsilon = \varepsilon_0\left[1 - \frac{\omega_p^2}{\omega^2}\right] \tag{12}$

The ω_p is plasma frequency and $\hbar\omega$ is located in visible or ultraviolet photon energy region. Below this energy level, real metal behaves like an ideal metal and its optical reflectivity is close to 1. On the other hand, above this energy regime, free carriers in metal cannot follow the variation of electric field and thus electromagnetic wave can penetrate and propagate freely into metal. If $\omega = \omega_p$, $\varepsilon = 0$ according to Eq. (12) and wavelength of light becomes infinite. This resonance condition corresponds to a collective excitation called plasmon [1–4]. In other words, plasmon oscillation or coupling of electromagnetic field with metals and semiconductors can affect their electron subsystem [5–10]. When external field is changed, the electron subsystem can oscillate in the field leading to surface charges along the field.

On the other hand, optical second harmonic generation (SHG) appears to be one of the most attractive as a tool for metal surface study due to a main reason of being highly surface-specific for any solid state materials [11–15]. When electron is in ideally parabolic potential, it makes a linear optical response to external electric field. In the contrary, if electron is in non-parabolic potential, its response to the external field is nonlinear. In nonlinear optical effect, magnitude of generated polarization is not proportional to that of incident electric field.

Nonlinear polarization as a function of incident electric field can be written as

$$\vec{P} = \varepsilon_0 \left[\chi^1 \vec{E}^1 + \chi^2 \vec{E}^2 + \chi^3 \vec{E}^3 + ... \right] \tag{13}$$

$$\vec{P} = \vec{P}^1 + \vec{P}^2 \vec{P}^2 + ... \tag{14}$$

$$\vec{P} = \vec{P}^1 + \vec{P}^{NL} \tag{15}$$

The χ^2 and χ^3 are nonlinear susceptibility tensors, whereas $\vec{P}^{NL}$ is nonlinear polarization.

The ith component of second order nonlinear polarization is

$$P_i^2(\omega_i) = \sum_{jk} \varepsilon_0 \chi_{ijk}^2 (-\omega_i, \omega_j, \omega_k) E_j(\omega_j) E_k(\omega_k) \tag{16}$$

where $\vec{P}(\omega_i) \propto e^{-i\omega_i t}$ and $\vec{E}(\omega_j) \propto e^{-i\omega_j t}$.

The nonlinear optical response of material enables us to perform a surface sensitive spectroscopy including optical second harmonic spectroscopy of solid surfaces. Second-order process, such as SHG ($\omega_2 = 2\omega_1$), is a phenomenon in which polarization

$$P = \chi^2 E^2 \tag{17}$$

induced by incident electric field $E \propto \exp(-i\omega t)$, radiates a light of double frequency $E \propto \exp(-2i\omega t)$. The SHG occurs in a medium whose structure lacks inversion symmetry like structures near nanowire surfaces as shown in Fig. 1.

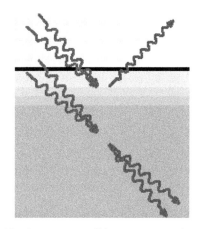

Figure 1 The double photon energy (blue wavy arrows) produced by collision of single-photon energy (red wavy arrows).

Since the second-order responses of our samples depend sensitively on the detailed structure of the samples, they result in their nonlinear responses [12,13]. In order to see how sensitive the SHG technique is, this article focuses on SHG as a noninvasive monitor of plasmon oscillation in metallic nanowires.

Theoretical Model and Analysis

This article presents a simple model for calculation of absorption spectra, Maxwell–Garnett model [16] in which objects of short wire axes b are assumed to be smaller than wavelength of incident light (Fig. 2). By making aspect ratio of a/b to be infinity, one can treat ellipsoid-shape metal as metallic nanowire.

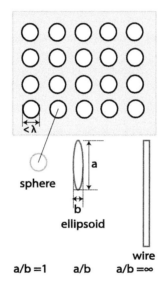

Figure 2 Composite particles in host medium. The a and b are long and short axes of metallic nanowires, respectively.

In this quasi-static regime, we are able to calculate the field distributions to metallic nanowires placed in an external uniform field. In metallic nanowires, the applied static electric field induces a dipole moment and polarizability α is therefore defined through [1,2]:

$$\alpha = \frac{V}{3Q} \left[\frac{\varepsilon_m(\omega_p) - \varepsilon_0}{\varepsilon_m(\omega_p) + (Q^{-1} - 1)\varepsilon_0} \right] \qquad (18)$$

where ε_m and ε_o are dielectric constants of metallic nanowires and environment, respectively, ω_p denotes resonance frequency, V is volume fraction, and Q is depolarization factor depending on ratio of long wire axes a/short wire axes b. The electric polarizability α in Eq. (18) can show a resonance behavior when a real part of denominator $\mathrm{Re}\{\varepsilon_m(\omega_p) + (Q^1 - 1)\varepsilon_o\}$ becomes zero, which is a case for $\mathrm{Re}\{\varepsilon_m(\omega_p)\}$ $= \varepsilon_o - \varepsilon_o Q^{-1}$.

From Eq. (18), surface plasmon mode of metallic nanowires is dependent on dielectric constant ε, resonance frequency ω_p, and depolarization factor Q. First, when depolarization factor Q is in between 0 and 1, surface plasmon excitation exists because their real part $\mathrm{Re}\{\varepsilon_m(\omega_p)\}$ fulfills resonance condition $\mathrm{Re}\{\varepsilon_m(\omega_p) + (Q^{-1} - 1)$ $\varepsilon_o\}$ at visible light. Second, resonance frequency ω_p is dependent on depolarization factor Q and dielectric constant of environment ε_o. Then, one can tune plasmon excitation via Q and ε_o. That is, when Q and ε_o are enhanced, plasmon maxima tend to be red-shift. A defined tuning of ω_p over a whole visible spectrum region is then possible. Last, imaginary part of $\varepsilon(\omega)$ can predict a tendency of spectral width of surface plasmon profile. When imaginary part of $\varepsilon(\omega)$ is reduced (Table 1 and Fig. 3), the profile becomes sharp.

Table 1 Dielectric constant of different metals

Metals	ε_m (λ = 1064 nm)	ε_m (λ = 532 nm)
Silver	$-58.14 + i0.61$	$-11.78 + i0.37$
Gold	$-48.24 + i3.59$	$-4.71 + i2.42$
Copper	$-49.13 + i4.91$	$-5.50 + i5.76$

Since the electrostatic application as explained in Maxwell–Garnet model accounts for a homogeneous distribution of electromagnetic field entirely the metallic wire volume, the more realistic methods, namely, finite-difference time-domain method (FDTD), need to be applied [17,18]. The FDTD calculation can solve an entire set of Maxwell's equations for coupling of light with any shape and size of metallic nanowires. Maxwell's curl equations accompanied with dielectric function of metals by Drude model are generally valid for dielectric materials, and their electromagnetic fields are able to solve simultaneously in time and space.

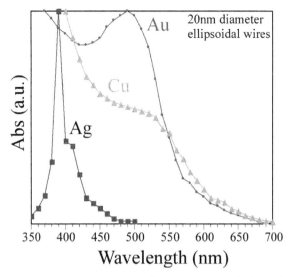

Figure 3 Absorption spectra of metallic nanowires calculated by finite-difference time-domain method.

For practical application of FDTD, the commercial software developed by Lumerical Inc. from Canada is exploited [19]. This program simulates different conditions for coupling of light with metallic nanowire shape and size and gives outcome of their absorption spectra and field distributions. Input parameters for FDTD programs are: shape of object, material of interest, cell size, and incident angle of incident light. Advantage of FDTD technique is that we can use very tiny cubic cells to cover all surfaces of interest. This guarantees an accuracy of field distribution over metallic nanowires. A drawback of this technique is that calculation time rises with increasing the number of cells. For example, one simulation time is approximately 500 fs as shown in Fig. 4. In addition, the fine discretization in finite differencing increases the rounding off error as well. Hence, a balance between accuracy and round off error needs to be maintained in FDTD method.

The simulation has converged when the component of electric field has decayed to zero before the end of the simulation time. The intensity of the total field is calculated and normalized to that of the incident field. For all simulations, absorbing boundary conditions based on the perfectly matched layer approximation are applied.

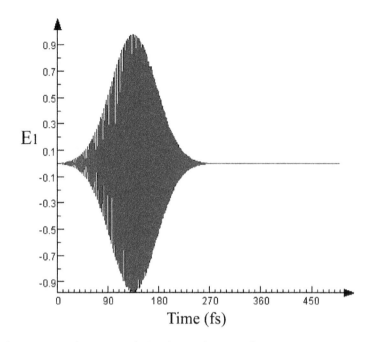

Figure 4   FDTD's main simulation time is about 500 fs.

Fig. 5 shows a model of metallic nanowires used as object for FDTD calculation. The nanowires are surrounded by 2-D unit cell with a grid size of 0.1 nm × 0.1 nm. The geometrical parameters for the long wire axes (a) is 200 nm and for the short axes (b) is 20 nm. These calculated parameters provide an aspect ratio (a/b) of 10 corresponding well with the aspect ratio obtained from our previous experimental data of fabricated nanowires [9]. By considering x-polarized light traveling in the forward y-direction and x-polarized light traveling in the forward y-direction, the three electromagnetic field components $E_x(x,y,t)$, $E_y(x,y,t)$, and $H_z(x,y,t)$ for each polarization direction can be calculated. Calculated absorption spectra agree well with typical experimentally obtained absorption spectra partly due to surface plasmon modes [9]. With increasing the short wire axes b at the same aspect ratio a/b, the absorption spectra shift to lower photon energy region. This is in agreement with the fact that retardation effect, which means external non-uniform field around metallic nanowires, has significant influence on plasmon maxima.

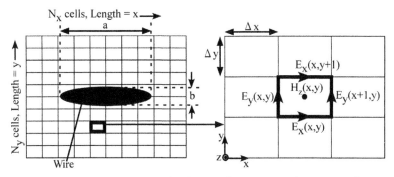

Figure 5 Coordination system for finite-difference time-domain simulation. The *a* and *b* are long and short wire axes of metallic nanowires, respectively. The aspect ratio is *a/b* related to geometrical effecting on the surface plasmon mode.

On the other hand, surface plasmon mode leads not only to a consequence for the absorption spectra, but also to field enhancement in second harmonic generation (SHG) process induced by a broken symmetry. Let us consider a potential of electron of mass m along the coordinate z,

$$\frac{V}{m}(z) = 0.5\omega_0^2 z^2 + a_3 z^3 + a_4 z^4 + \dots \tag{19}$$

If this potential is symmetric with respect to the point $z = 0$, we have $a_{2n+1} = 0$. In contrast, if it is asymmetric, the equation of motion of the electron will be

$$m\frac{d^2 z}{dt^2} + m\omega_0^2 z + 3a_3 z^2 + 4a_4 z^3 + \dots = Fe^{-i\omega t} \tag{20}$$

The solution to this equation to the first order by the perturbation theory is

$$Z = Z^{(0)} + Z^{(1)} \tag{21}$$

where $\quad z^{(0)} = \dfrac{F}{m(\omega_0^2 - \omega^2)} e^{-i\omega t} \tag{22}$

and

$$z^{(1)} = \frac{-3a_3}{m(\omega_0^2 - 4\omega^2)} \frac{F^2}{m^2(\omega_0^2 - \omega^2)^2} e^{-2i\omega t} \tag{23}$$

Therefore, this proves that SHG is a phenomenon in which a polarization induced by incident electric field $E \propto \exp(-i\omega t)$ radiates a light of double frequency $E \propto \exp(-2i\omega t)$.

In terms of the field enhancement in second harmonic generation (SHG) process, it can be defined by [14,15]:

$$L = \left|\vec{E}\right| / \left|\vec{E}_o\right| = L(\vec{r}, \omega) \tag{24}$$

where $\vec{E}$ denotes optical field and $\vec{E}_o$ denotes incoming field. In Eq. (24), the field enhancement L is mainly determined by the field at position $\vec{r}$ and at resonance frequency ω_p of the exciting light. This expression depends on polarization conditions with respect to metallic nanowire surfaces as well. For polarization parallel to the nanowire axes, the external field is continuous across their surfaces. Then the enhancement is possible for a resonant excitation of the surface plasmon. For polarization perpendicular to the nanowire axes, the enhancement is also possible due to contribution of corner effect around their edges as shown in Fig. 6. The field near circular wires in Fig. 6 (top), which are highly symmetric, is homogeneous on the entire surface. A strong field in the single wires can be related to oscillating polarization charges with positive charges on the one side of the wires and negative charges on the other side. Polarization charges of opposite signs are confined between gap and they result in the electric field enhancements.

The polarization charge distributions for non-regular wires, which are less symmetric, are different from the circular ones. The charges are mainly concentrated around the corners. Namely, on the left side of the single square wires in Fig. 6 (middle), the charge distributions with the same sign (minus charges) are accumulated at the top (one minus charge at top left) and bottom (another minus charge at bottom left) of the corners. On the other hand, plus charges at the upper and lower corners are located on the right side of the single wires. These charges topology of the opposite signs (positive and negative charges) accumulated around the corners lead to a dipole-like field distribution between gap and result in the strong electric fields. Reducing the wire symmetry from square wires in Fig. 6 (middle) to triangular wires in Fig. 6 (bottom) increases the sharpness of corners, leading to a strongly confined charge and a dipolar interaction. The sharper the corner, the more confined the surface charges and the stronger the resulting field enhancements.

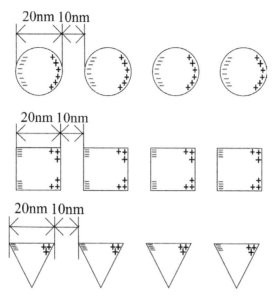

Figure 6 Field distributions of different cross-sectional shapes from metallic nanowires.

The electric field enhancements were found to depend on the asymmetric structure of the wires especially for noncircular shapes (triangular and square wires). Therefore, we suggest that the lightning-rod effect must be taken into account for an accurate description of the field distribution in real nanowire structures.

Electric field enhancement on SHG process also can be demonstrated by FDTD simulation via plasmon maxima in metallic nanowires [20,21]. Field enhancement exists near metallic nanowires for parallel polarization conditions. The electric field enhancement is dominant when a periodicity of the metallic nanowires is reduced. Two contributions existed according to the decrease of the periodicity. In short distance, short-range interactions between neighboring wires induce near-field coupling and create highly sensitive plasmons confined to metal boundaries. In contrast, when the periodicity exceeds the range of near-field coupling, far-field interactions prevail among wires. This mechanism can be explained by using a dipole–dipole interaction model. Dipole field in an individual wire induces the dipoles in the neighboring wires by distorting the neighbor's electron cloud and it leads to the formation of localized surface plasmons and then results in local electric field enhancement.

Conclusion

This article explains surface plasmon modes in general and concepts of second harmonic generation. A simple analytical method based on Maxwell–Garnett model is then presented for calculation of absorption spectra. A numerical method, finite-difference time-domain, is then shown and this allows one to accurately calculate plasmon maxima in metallic nanowires. It is found that Maxwell–Garnett simulation does not fit with experimental results due to wire sizes much larger wavelength of incident light; however, finite-difference time-domain calculation is in agreement well with the findings because finite-difference time-domain calculation can solve an entire set of Maxwell's equations for coupling of light with any shape and size of metallic nanowires.

References

[1] Gaponenko, S. V., *Introduction to Nanophotonics*, Cambridge, New York, pp. 130–140 (2010).

[2] Jahns, J., and Helfert, S., *Introduction to Micro- and Nanooptics*, Wiley, New York, pp. 215–229 (2012).

[3] Zangwill, A., *Physics at Surfaces*, Cambridge, New York, pp. 25–100 (2008).

[4] Samorjai, G., *Chemistry in Two Dimensions*, Ithaca, New York, pp. 50–125 (2011).

[5] Gutierrez, F. A., Salas, C., and Jouin, H., Bulk plasmon induced ion neutralization near metal surfaces, *Surface Science*, 2012, 606 (15–16), pp. 1293–1297.

[6] Yeshchenko, O. A., Bondarchuk, I. S., Dmitruk, I. M., and Kotko, A. V., Temperature dependence of surface plasmon resonance in gold nanoparticle, *Surface Science*, 2013, 608, pp. 275–281.

[7] Ning, J., Nagata, K., Ainai, A., Hasegawa, H., and Kano, H., Detection of influenza virus with specific subtype by using localized surface plasmons excited on a flat metal surface, *Japanese Journal of Applied Physics*, 2013, 52, pp. 82402–82405.

[8] Mott, D., Lee, J. D., Thuy, N., Aoki, Y., Singh, P., and Maenosono, S., A study on the plasmonic properties of silver core gold shell nanoparticles, *Japanese Journal of Applied Physics*, 2011, 50, pp. 65004–65011.

[9] Locharoenrat, K., Sano, H., and Mizutani, G., Phenomenological studies of optical properties of Cu nanowires, *Science and Technology of Advanced Materials*, 2007, 8, pp. 277–281.

[10] Wunderlich, S., and Peschel, U., Plasmonic enhancement of second harmonic generation on metal coated nanoparticles, *Optic Express*, 2013, 21 (16), pp. 18611–18623.

[11] Dounce, S. M., Yang, M., and Dai. H. L., Physisorption on metal surface probed by surface state resonant second harmonic generation, *Surface Science*, 2004, 565 (1), pp. 27–36.

[12] Locharoenrat, K., Sano, H., and Mizutani, G., Rotational anisotropy in second harmonic intensity from copper nanowire arrays on the NaCl (110) substrates, *Journal of Luminescence*, 2008, 128, pp. 824–827.

[13] Locharoenrat, K., Sano, H., and Mizutani, G., Second harmonic spectroscopy of copper nanowire arrays of on the (110) faceted faces of NaCl crystals, *Journal of Physics: Conference Series*, 2008, 100, pp. 52050–52053.

[14] Bloembergen, N., *Nonlinear Optics*, Wiley, New York, pp. 27–38 (2005).

[15] Shen, Y. R., *The Principle of Nonlinear Optics*, Wiley, New York, pp. 3–32 (2004).

[16] Garnett, J. C. M., Colours in metal glasses and in metallic films, *Philosophical Transactions of the Royal Society of London A*, 1904, 203, pp. 385–420.

[17] Taflove, A., *Computational Electrodynamics: The Finite-Difference Time-Domain Method*, Artech House, Boston, pp. 95–115 (2005).

[18] Sullivan, D. M., *Electromagnetic Simulation using the FDTD Method*, IEEE, New York, pp. 42–77 (2005).

[19] FDTD Solutions v8.7.4, https://www.lumerical.com/downloads/

[20] Locharoenrat, K., and Mizutani, G., Characterization, optical and theoretical investigation of arrays of the metallic nanowires fabricated by a shadow deposition method, *Advanced Materials Research*, 2013, 652, pp. 622–623.

[21] Locharoenrat, K., Sano, H., and Mizutani, G., Field enhancement in arrays of copper nanowires investigated by the finite-difference time-domain method, *Surface and Interface Analysis*, 2008, 40, pp. 1635–1638.

Index